METROPOLITAN DEMOCRACIES

Metropolitan Democracies

Transformations of the State and Urban Policy in Canada, France and Great Britain

Edited by

PHILIP BOOTH
University of Sheffield

BERNARD JOUVE
Université du Québec à Montréal

Routledge
Taylor & Francis Group

LONDON AND NEW YORK

First published 2005 by Ashgate Publishing

Reissued 2019 by Routledge
2 Park Square, Milton Park, Abingdon, Oxon, OX14 4RN
52 Vanderbilt Avenue, New York, NY 10017

Routledge is an imprint of the Taylor & Francis Group, an informa business

A Library of Congress record exists under LC control number:

ISBN 13: 978-0-8153-9051-0 (hbk)
ISBN 13: 978-1-138-35651-1 (pbk)
ISBN 13: 978-1-351-15308-9 (ebk)

Contents

List of Figures

List of Tables

List of Contributors

Philip Booth is Reader in Town and Regional Planning at the University of Sheffield. His work has focused on partnership in urban policy and on planning in a comparative perspective. Among his most recent publications are: 'From Property Rights to Public Control: the Quest for Public Interest in the Control of Development', *Town Planning Review* **73**(2), pp. 153-170; *Planning by Consent: the Origins and Nature of British Development Control*, London, Routledge; 'Promoting Radical Change: the *Loi relative à la solidarité et au renouvellement urbains* in France', *European Planning Studies* **11**(8), pp.949-963. Contact: p.booth@sheffield.ac.uk

Julie-Anne Boudreau is Professor in the Department of Political Science at York University, Toronto. Her research has taken her to Los Angeles, Montréal and Toronto, towns in which she has studied social movements, local democracy and forms of government. She has published *The Mega-City Saga: Democracy and Citizenship in this Global Age*, Black Rose Books, 2000. She is member of a research group on urban fragmentation in various cities in the world, a project that is led by Jaglin and Coutard and financed by the CNRS. In addition, she has begun a research project funded by the CRSH whose title is 'Metropolitan Governance and International Competitiveness: the Example of Montréal and Toronto'. Contact: jab@yorku.ca

Raphaël Canet holds a doctorate in sociology from the Université du Québec au Montréal (2002) and has been coordinator of the Canadian research chair in globalisation, citizenship and democracy at UQAM since 2001. His current research is concerned with theories of the state and nationalism, the question of identity of political mobilisation, and the transformation of citizenship in the context of globalisation, the information society and governance. He has, in collaboration with Jules Duchastel, published *La nation en débat. Entre modernité et postmodernité*, Editions Athéna, 2003. Contact: canet.raphael@uqam.ca

Didier Chabanet holds a doctorate in political science and is currently responsible for research at Inrets (Institut national de recherche et d'évaluation sur les transports et leur sécurité), research associate with the Groupe d'analyse des politiques publiques at the Ecole Normale de Cachan, and member of Ceriep (Centre de politologie de Lyon). He is particularly interested in movement generated by the European Union, in collective action by young second generation immigrants in France, and more recently, public action on road safety. Among his recent publications are: *L'action collective en Europe/Collective Action in Europe* (edited with Richard Balme and Vincent Wright), Paris, Presses des Sciences Po, 2002. Contact: didierchabanet@hotmail.com

Gordon Dabinett is Reader in Town and Regional Planning at the University of Sheffield. His work is particularly concerned with urban regeneration policy. He has been member of the New Deal for Communities working party. Among his most recent publications are: (2002) 'Reflections on Regional Development Policies in the Information Society', *Planning Theory and Practice* **3**(2), pp. 232-237; (2001) *A Review of the Evidence Base for Regeneration Policy and Practice*, a report for the Department of Environment, Transport and the Regions (with P. Lawless, J. Rhodes and P. Tyler); (1999) 'Urban Policy in Sheffield: Regeneration, Partnerships and People' in R. Imrie and H. Thomas (eds) *British Urban Policy: an Evaluation of the Urban Development Corporations*, London, Sage, pp. 168-185. Contact: g.e.dabinett@sheffield.ac.uk

Jules Duchastel is holder of a professorship in the Department of Sociology at the Université du Québec à Montréal and also holds the Canadian research chair in globalisation, citizenship and democracy. His research is concerned essentially with the analysis of new forms of political regulation in the context of the growing influence of international organisations and of the development of trans-national decision making. His current research is a continuation of previous work on the transformation of national political institutions in the history of Canada and of Québec since the 1940s, through a study of political discourse. He has published with Editions Athéna, Montréal, *Fédéralismes et mondialisation. L'avenir de la citoyenneté et de la démocratie* (2003). Contact: duchastel.jules@uqam.ca

Jean-Marc Fontan is Professor of sociology at the Université du Québec à Montréal and is a member of the Centre de recherches sur les innovations sociales dans l'économie sociale, les entreprises et les syndicats (CRISES). He is a specialist in the field of economic anthropology and the sociology of development. His work has been tied principally to means of development in the setting of metropolitan Montréal. Among his latest work is to be found: *Social Economy, International Debates and Perspectives*, edited with E. Shragge, Montréal, Black Roses Books (2000); *Entre la métropolisation et le village global* (edited, with J.-M. Fontan, J.-L. Klein, D.G. Tremblay, Presses de l'Université du Québec (1999). Contact: fontan.jean-marc@uqam.ca

Pierre Hamel is Professor in the Department of Sociology at the Université du Québec à Montréal. His research is concerned with social movements and their institutionalisation, and also with questions related to planning and urban development. His recent publications include: 'Urban Issues and the New Policy Challenges: the Example of Public Consultation Policy in Montreal' in C. Andrew, K.A. Graham, and S.D. Phillips (eds) *Urban Affaires Back on the Policy Agenda*, Montréal and Kingston, McGill-Queens University Press, pp. 221-238; Hamel, P. (2001) 'Enjeux métropolitains: les nouveaux défis' *International Journal of Canadian Studies*, **24**, pp. 105-127. Contact: pierre.hamel@unmontreal.ca

Melody Houk is currently a doctoral student in sociology at the Centre de sociologie des organisations (CNRS/FNSP) in Paris. Her thesis topic focuses on the emergence of a tier of local government below that of the municipality in the three biggest French cities, Paris, Marseille and Lyon, since the beginning of the 1980s. She was a member of the monitoring unit for local democracy in the 20th arrondissement of Paris from 1997 to 2001. In 2001, she published 'Vers une décentralisation municipale à Paris?' in the review *Esprit* **6**, pp. 193-200. Contact: m.houk@cso.cnrs.fr

Bernard Jouve is holder of the Canadian research chair in the study of territorial dynamics. He is also assistant professor in the Department of Geography at the Université du Québec à Montréal and a member of the Centre des recherches sur les innovations sociales dans l'économie sociale, les entreprises et les syndicats (CRISES). He has worked for several years on urban governance in a comparative perspective. His recent work includes: (2004) *Horizons métropolitains. Politiques et projets urbains en Europe.* (ed. with C. Lefèvre), Lausanne, PPUR; (2003) *Les politiques de déplacements urbains L'innovation en question en cinq villes européennes,* Paris, L'Harmattan; (2003) *La gouvernance urbaine en questions,* Paris, Elsevier; 'Gouvernance métropolitaine: vers un programme de recherche comparatif', *Politiques et sociétés,* **22**(1), pp. 119-142. Contact: jouve.bernard@uqam.ca

Juan-Luis Klein is Professor in the Department of Geography at the Université du Québec à Montréal and a member of the Centre des recherches sur les innovations sociales dans l'économie sociale, les entreprises et les syndicats (CRISES), of which he became deputy director in June 2003. His work is concerned with on local and regional development, on economic restructuring and on the location of collective action. He has produced various books including (2003) *Reconversion économique et développement territorial* (with J.-M. Fontan and B. Lévesque), Presses de l'Université du Québec. Articles include: (2003) 'Reconversión y desarrollo a través de la iniciativa local: el caso de Montreal en Quebec', *Revista latinoamericana de estudios urbanos y regionales* **XXIX**(86), pp. 69-88 (with J.-M. Fontan and D.-G. Tremblay); (2003) 'Systèmes productifs locaux et réseaux productifs dans la reconversion économique: le cas de Montréal', *Géographie, Economie, Société* **5**(1) pp.59-75 (with D.-G. Tremblay and J.-M. Fontan). Contact: klein.juan-luis@uqam.ca

Anne Latendresse is Assistant Professor in the Department of Geography at the Université du Québec à Montréal and member of the Centre des recherches sur les innovations sociales dans l'économie sociale, les entreprises et les syndicats (CRISES). She is currently undertaking work on public consultation and participation in the *arrondissements* of Montréal since municipal reform. She is interested in the practices and strategies of urban movements, in questions of participative democracy and territorial governance. Her recent works include: (2003) 'Le local comme nouvelle scène de gouvernance et de développement à Montréal et a Montevideo' (with J.-L. Klein, J.-M. Fontan and M.-P. Paquin-Boutin) *Géographies et cultures* **45**, pp. 57-72; (2002) 'Réorganisation municipale sur l'île de Montréal: une opportunité pour la démocratie montréalaise?' *Annales des Ponts et Chaussée* 102, pp. 23-31. Contact latendresse.anne@uqam.ca

Benoît Lévesque has been Professor in the Department of Sociology at the Université du Québec à Montréal since 1982. He is also chair of the scientific committee of the Centre interdisciplinaire de recherche et d'information sur les entreprises collectives (CIRIEC). He was president of CIRIEC-Canada from 1995 to 2001. He has been a member of the Centre des recherches sur les innovations sociales dans l'économie sociale, les entreprises et les syndicats (CRISES) since 1990. Included in his recent publications are: (2000) *La création d'entreprises par les chômeurs et les sans-emploi: le rôle de la microfinance,* Montréal, CRISES-Université Concordia; (2000)

Le fonds de solidarité des travailleurs du Québec (FTQ): nouvelle gouvernance et capital de développement, Montréal, CRISES; (1999) 'La réingénierie des services financiers: un secteur exemplaire de l'économie des services. Le cas des Caisses populaires et d'économie Desjardins', *Lien social et Politiques*, 40, pp.89-103. Contact: levesque.benoit@uqam.ca

Peter Newman is Senior Lecturer in the School of Architecture and the Built Environment at the University of Westminster. His interests focus particularly on planning and urban governance in Europe. His recent publications include: (2002) *Governance of Europe's City Regions*, London, Routledge; (2000) 'Changing Patterns of Regional Governance in the European Union' *Urban Studies*, 37(5-6). Contact: newmanp@westminster.ac.uk

Andy Thornley is Reader in Urban Planning Studies at the London School of Economics and Political Science. He is interested in the link between planning and urban policy, and particularly in major urban projects seen in a comparative perspective. Among other works, he has published (1993) *Planning under Thatcherism*, London, Routledge; (1996) *Urban Planning in Europe: International Competition, National Systems and Planning Projects* (with P. Newman), London, Routledge; and (2002) *Metropolitan Governance and Spatial Planning* (with A. Kreukels), London, Spon. His latest work (2004), with P. Newman is entitled *Planning World Cities: Globalisation, Urban Governance and Policy Dilemmas*. Contact: a.thornley@lse.ac.uk

Jean-Yves Toussaint is Professor in the research laboratory 'Equipe développement urbain' (UMR CNRS 5600 Environnement, Ville, Société) attached to the Institut national des sciences appliquées in Lyon. His research focuses on the design of technical infrastructure in an urban setting, and the way in which practice and use are integrated into such infrastructure. Among his recent works are: (2003) *Pratiques techniques, pratiques démocratiques. La mise en œuvre de la démocratie*, Lausanne, Presses polytechniques et universitaires romandes, Coll. des Sciences appliquées (ed., with Monique Zimmermann); (2001) *User, observer, programmer et fabriquer l'espace public. Réflexion autour de l'expérience lyonnaise*, Lausanne, Presses polytechniques et universitaires romandes, Coll. des Sciences appliquées (ed., with Monique Zimmermann). Contact: toussaint@insa-lyon.fr

Sophie Vareilles is a doctoral student in geography, development and town planning in the research group 'Equipe Développement urbain' (UMR CNRS 5600 Environnement, Ville, Société) attached to the Institut national des sciences appliquées in Lyon. Her research focuses on the arrangements for participation in urban projects. Her recent publications include: (2004) 'Le projet urbain: espaces publics et pratiques de concertation: l'exemple de Lyon' in M. Zepf (ed.) *Action publique et métropolisation: concerter, gouverner et concevoir les espaces publics urbains*, Lausanne, Presses polytechniques et universitaires romandes (with J.-Y. Toussaint, Monique Zimmermann). Contact: sophie.vareilles@insa-lyon.fr

Philippe Warin is Director of Research at the Centre national de recherche scientifique and lectures at the Institut d'études politiques at Grenoble where he directs the Centre de recherche sur la politique, l'administration, la ville et la territoire (CERAT). His programme of research is concerned with citizenship in the implementation of public policy, that he approaches through the analysis of breakages in public supply of services, of the openness of public administration to social demand and of citizen participation in the production of public action. His publications include: (2002) 'The Role of Nonprofit Associations in Combatting Social Exclusion in France', *Public Administration and Development* 22, pp. 73-82; (2002) *Les dépanneurs de justice. Les 'petits fonctionnaires' entre qualité et équité*, Paris, Librairie générale de droit et de jurisprudence. Contact: philippe.warin@upmf.grenoble.fr

Marcus Zepf has a degree in architecture from the Technische Universität München and a doctorate in technical science from the Ecole polytechnique fédérale de Lausanne. He is currently research associate in the research group 'Equipe Développement urbain (CNRS UMR 5600 Environnement, Ville, Société) attached to the Institut national des sciences appliquées at Lyon. Recent publications include: (2004) *Action publique et métropolisation: concerter, gouverner et concevoir les espaces publics urbains* (ed.), Lausanne, Presses polytechniques et universitaires romandes; (2001) *Weichenstellungen im Ländlichen Raum für Landwirtschaft und Stadt-Land-Beziehungen: Disput, Akteure, Wege*, München, Bayerische Akademie Ländlicher Raum; (2000) 'Using and Conceiving Public Space Influenced by Urban Transformation' in J. Benson and M.H. Roe (eds) *Urban Lifestyles: Spaces, Places, People* (with M. Bassand), Rotterdam, Balkema, pp. 57-64. Contact: zepf@insa-lyon.fr

Monique Zimmermann is Professor in the research group 'Equipe Développement urbain (CNRS UMR 5600 Environnement, Ville, Société) attached to the Institut national des sciences appliquées in Lyon. Her research focuses on the design of technical infrastructure in an urban setting, and the way in which practice and use are integrated into such infrastructure. Among her recent publications are: (2003) *Pratiques techniques, pratiques démocratiques. La mise en œuvre de la démocratie* (ed., with J.-Y. Toussaint), Lausanne, Presses polytechniques et universitaires romandes, Coll. des Sciences appliquées; (2001) *User, observer, programmer et fabriquer l'espace public. Réflexion autour de l'expérience lyonnaise* (ed., with J.-Y. Toussaint), Lausanne, Presses polytechniques et universitaires romandes, Coll. des sciences appliquées. Contact; gcudu@insa-lyon.fr

Preface

The origins of this book lie in a seminar of the French and British Planning Study Group held in Sheffield in September 2002, which was supported financially by the French Embassy in London, the University of Sheffield's Department of Town and Regional Planning, the Université du Québec à Montréal and the Research Chairs of Canada programme. The translation of some of the chapters was made possible by virtue of financial support from the Faculté des Sciences Humaines and the Centre de Recherche sur les Innovations Sociales at the Université du Québec à Montréal. To all of these our thanks are due. The book first appeared in French under the title *Démocraties metropolitaines: transformations de l'Etat et politiques urbaines au Canda, en France et en Grande-Bretagne,* published by the Presses de l'Université du Québec in March 2004.

The maps both here and in the Canadian edition were prepared by André Parent, cartographic technician at the Department of Geography at the Université du Québec à Montréal and by Paul Coles at the Department of Geography at the University of Sheffield. Dale Shaw and Paul O'Hare at the University of Sheffield worked to produce and sub-edit the English edition. Their work has been invaluable to the successful completion of the book.

Philip Booth
Bernard Jouve

Sheffield/Montréal

Chapter 1

Metropolitan Cities at the Crossroads of Globalisation and Changing Politics

Bernard Jouve

This book stems from a seminar held by the French and British Planning Study Group at the University of Sheffield in September 2002, organised by Philip Booth. The theme of the seminar was 'Acting together in urban regeneration in France and Great Britain'. It quickly became apparent that this theme transcends the question of urban planning itself, referring to a more general set of issues related to the organisation of politics in modern democracies, the renewed practice of political participation, and the foundations of citizenship. We thus decided to expand on the general set of themes, increase the number of contributors, and broaden the framework of comparison by including Canada. This decision can be explained by the vitality of participatory democracy in cities such as Montreal and Toronto, but also by the fact that Canada, in developing a policy of integration based on multiculturalism, has become a model no longer regarded as 'exotic,' but one that is brought up more and more often in debate.

Indeed, due to the extent of migratory flows between developing countries and the developed world, the emergence of sub-national identity movements based on nationalist sentiments and the rising power of community-based social relations, 'the Canadian way' is being raised more and more in discussion, and appears to some as the best solution, in terms of both effectiveness and justice, enabling the integration/differentiation dialectic within pluralistic societies to be addressed (Kymlicka 1998). Metropolitan cities and the policies that are developing within them, present an opportunity to observe the 'Canadian model'. This question of communitarianism is one of the main factors behind the comparison. The chapters that follow address these questions: What are the dynamics common to all three states which are currently changing the practice of democracy and the definition of citizenship in their metropolitan cities? Aside from differences stemming from their various histories in the process of becoming nation-states, the values that give them shape, and their political and administrative cultures, can a process of convergence towards the 'Canadian model' be observed? The choice of Britain, France and Canada as case countries is one based on the fact that they represent a continuum between two major approaches to citizenship. The 'single and indivisible' universalistic approach is exemplified within the French Republic, and the Canadian category-based approach which recognises and facilitates the expression of cultural, linguistic, ethnic and religious differences. Great Britain currently occupies the middle ground between these two poles.

A Series of Questions

This book also addresses another series of questions related to the evolution of intergovernmental relations and the governance of metropolitan cities. In Canada, the influence of the federal government in urban policies is quite limited because of the way jurisdictions are distributed among different levels of government, with 'municipal affairs' being the prerogative of provincial governments. This division of labour may soon be changed. Paul Martin, current prime minister of Canada, has made clear his intention to develop a strong partnership between the federal government and the municipalities, particularly the metropolitan cities. This proposal will no doubt raise the ire of the provinces, ever inclined to see municipalities as their own 'creations'. In France and Great Britain, over the last twenty years or so, urban policy has been a vehicle for the reconstruction of intergovernmental relations, due to the opposing logic and rationality of each level of institution. The main reforms, launched by the governments of Margaret Thatcher and extended by Tony Blair, resulted in the centralising of intergovernmental relations to the advantage of the central government. Great Britain, home of Local Government, saw the political influence of local communities, as well as that of democratically elected urban institutions, diminish. This change took place to the benefit of economic actors that were heavily involved in developing urban policies through new institutionalised forms of public/private partnership within new structures named Quasi-Autonomous Non-Governmental Organisations (QANGOs). This policy has been progressively modified, first by the government of John Major and then by Tony Blair's government, both increasing the number of 'partners' involved. Thus urban policies have gradually become a space policy in which the reconstruction of the relationships between civil society and politics can take place.

During the same period, France opted for a different institutional arrangement by means of decentralisation. From 1982 onwards, a series of laws succeeded one another, giving local governments more fields of jurisdiction and responsibilities. This transformation of government structure eventually led to a significant amendment of the Constitution in 2003, which now stipulates that France is a decentralised state.

Picking up on a classic investigation in political science which aims to determine the impact institutions and various kinds of legislative framework have on the content of public policies, it seemed pertinent to compare these three countries in order to answer, in part, the equally traditional question: Who governs the cities? In a general context characterised by an acute neo-liberal agenda, the reconstruction of states and the rise in power of metropolitan cities as essential spaces for new forms of network-based regulation that come with globalisation (Castells and Hall 1994; Castells 2000), how does this 'new' division of labour between central governments and cities play out? What degree of autonomy do they enjoy? How does globalisation, the reconstruction of social relations within metropolitan cities, and the transition towards a 'second modernity' (Beck 1992) affect the practice of democracy and redefine citizenship?

Finally, this book is intended to add to the body of literature, which has grown particularly abundant recently, on the transformation of politics in Western countries. Indeed, while the institutional framework in Canada, France and Great Britain may differ, the issue of the radical reform of the democratic relationship and of citizenship in action is equally high on the political agenda of cities in all three countries. This redefinition of the relationship between civil society and the political realm is certainly not new. Over the last 30 years, there has been a multitude of reports, conferences, books and articles addressing this 'political crisis', the widening gap between elected representatives and their constituents, and the inability of the 'political world' to fulfil the expectations of civil society. This diagnosis is put forward both by political essayists (Lamoureux 1999; Comor and Beyeler 2002; Courtemanche 2003) and academics (Norris 1999; Skocpol and Fiorina 1999; Pharr and Putnam 2000; Balme et al. 2003) all of whom point to the growing mistrust of politics by 'civil society'. It is precisely in this context that urban policies in the three countries examined here – but the sample group could easily be broadened – are currently taking on particular importance.

This mistrust towards the political realm is mainly expressed in relation to an organisational mode that is centred on the principle of political representation, and therefore on the central role played by elected representatives in the decision-making processes regarding resource allocation and arbitration of conflicts. Conversely, there are growing expectations in favour of a renewed democracy, based on dialogue, deliberation and partnership; in other words, on the pluralisation of the political system. The 1990s were thus characterised by a major phenomenon whereby the sub-national territory and the local level have become, for many observers, specialists and decision-makers, the alpha and omega of any overhaul of the political system. As transformation of the political order is deemed impossible through action at the central government level, it is felt that any effective intervention must take place at the level of local governments. The proximity of political and administrative decision-makers and citizens at the local level is believed to naturally facilitate innovation, change and the opening up of decision-making systems (Loughlin 2001). This theme is not new. It can be found as early as the 1960s and 1970s in many Western countries. In some cases, it has served to fuel radical criticism of the face of the modern state, subservient to the interests of capitalism and, above all, a source of indoctrination and alienation. The local level, then, came to be synonymous with 'small is beautiful', the counter culture and alternative self-management projects. These radical views have since disappeared. What remains is that the case for the state as an authority based on which a renewed political order can be built, giving more room to the 'citizen approach', to quote from the alternative society movement particularly in style these days, is closed. Neither conservatives, by ideology, nor liberals (in the North-American sense), nor the social democrats, from experience and by defiance, support this platform. Thus, sub-national political spaces and, in particular, metropolitan cities, are being turned to for solutions once again. Hence, the example of Porto Alegre and its participative budget is often raised to show that politics can be done differently including, and perhaps especially, with disadvantaged social groups (Gret and Sintomer 2002).

Lastly, other analysts see in the metropolitan cities spaces within which it is possible to implement mechanisms of adaptation to globalisation and advanced capitalism, and to control its most harmful effects in terms of exclusion and increasing insecurity. Based on the writings of K. Polanyi on the importance of local spaces in the 'great transformation' linked to the development of industrial capitalism in Great Britain (Polanyi 1944), many observers see the metropolitan city, and no longer the nation-state, as the level at which development models can be generated that succeed in combining both economic growth and social justice.

This book proposes, in both analytical and empirical terms, a scientific debate over what closely resembles a 'Neo-Tocquevillian mirage'. What is the real scope of renewal in participatory and deliberative democracy in British, Canadian and French metropolitan cities? Are dialogue and partnership between elected representatives and civil society (community and ethnic groups, economic actors) actually transforming the basis of the political order? This book also aims to address, in some measure, this second series of questions.

With these questions in mind, we approached experts at British, Canadian and French universities. Before briefly introducing them, a cautionary note is called for. This book is not intended to be a point-by-point comparison between Canada, France and Great Britain. That would have required designing an inclusive research program based on a precise methodological protocol, something that was turned down by all of the participants. We had neither the material resources nor the time. However, it is our hope that the arguments presented here will enable just such an international program to get underway, with other countries being integrated as well. The purpose of this book is rather to identify the thrust of such a program and to formalise a common research agenda regarding the complex and ambiguous relations currently developing between globalisation, the reconstruction of nation-states, the practice of local democracy and the transformation of citizenship in Western metropolitan cities.

Chapter Contents

It is precisely this issue of globalisation and its effects on the transformation of citizenship that is at the centre of the first two chapters of this book. Duchastel and Canet point out that the changes accompanying the globalisation in all manner of exchanges have the effect of calling into question a form of citizenship that increasingly seems outdated. Traditionally, citizenship was considered only in terms of the state being the main socio-political construct in the regulation of societies. Given that globalisation brings with it changes in territoriality, political regulation and citizenship, Duchastel and Canet are interested in the emergence of new forms of citizenship and have developed a typology of various forms of democracy: centralised representative, decentralised representative, supranational, corporate, protest-based, and radical. The heuristic nature of this typology lies mainly in the fact that it formalises a transformation of the political order in which territoriality plays a major role. Globalisation has meant that the term territoriality in modern societies is no longer solely centred on the state and its political, administrative and legislative bodies. The state is most certainly not extinct and its institutions continue to exert great influence over the relationships between sub-state actors. That said, alternative forms of practising democracy, no longer dependent on election-based representation, are emerging at new territorial levels. The strictly supranational level constitutes one of these new reference points with regard to politics in particular. This is both through integration processes such as the European Union and also within international forums such as the World Trade Organization or the World Social Forum (and its 'regional' variations in Africa, Europe and Asia, etc.). The local level is also the seat of this reconstruction with a democracy that is increasingly expressed in protest-based or even radical terms; and which calls into question the principle of authority associated with a state-centred model of political practice.

In his chapter, Hamel examines the transformation of democracy at the metropolitan level by focusing on the social, economic and political transformations that accompany metropolitanisation. Briefly reviewing the evolution of urban morphology, which is given concrete expression through a generalisation of the urban level, Hamel further emphasises the fact that the mode of political integration centred on the state and its institutions has run out of steam. Metropolisation is characterised by more pluralistic societies, where the question of recognising the distinctiveness of individuals and social groups structures is on the political agenda. Hamel argues, what is involved here is the transformation of a political order centred on the state and on the gradual assertion of metropolitan cities as new political territories within which a dual process of differentiation and social integration is taking place. Hamel sees in this current process historical conditions for the gradual assertion of a metropolitan-based citizenship in which both 'classic' institutions, having the legitimacy that stems from elections and urban social movements, will play an essential role. Far from opting for a mechanical interpretation of the causal relationship between globalisation and the assertion of metropolitan citizenship, Hamel stresses, conversely, that there are limits to the extent to which the current political order can be surpassed. Established institutions, structures of decision-making systems, positions of prominent members of the elite are all at stake. Hamel is understandably cautious regarding the real short and medium-term repercussions of the dynamics that are currently at play. He does, however, identify several themes that, in his opinion, constitute the vehicles for political reconstruction: poverty; social inequality; the integration of immigrants and cultural communities; and protection of the environment. Like Duchastel and Canet, Hamel in fact considers that globalisation is bringing about a change of scale in the regulation of modern societies; regulation that entails, on the one hand, mechanisms for resource allocation to individuals and social groups and, on the other hand, the resolution of conflicts between these actors, and, finally, the crystallisation of new identity-based relations. As such, metropolitan cities constitute a new level of collective action that is gradually emerging as an additional level among the global, national and local levels. The new division of labour does not so much come from a zero-sum gain in which what is 'gained' by one territorial level is 'lost' by another. Even in Western Europe, where the process of supranational integration has made most headway, there remain very few authors who support the thesis of the dissolution of the state. The 'end of territories' must instead be considered as the calling into question of a solely state-centred mode of doing politics (Badie 1995). We are presently witnessing a re-engineering of regulation systems and multi-level governance (Hooghe and Marks 2001) in which the state continues to play an essential role. The state does not dominate to the same degree as it once did; its 'grip has been loosened' (Le Galès 1999), but it has nonetheless not become obsolete.

The 'death of the nation-state' often appears to be greatly exaggerated (Anderson 1995).

This aspect is dealt with in further detail in the chapters written by Dabinett, Booth, Newman and Thornley, and Boudreau. Dabinett reviews in detail the evolution of urban policies in Great Britain over the last twenty years. He demonstrates that urban policies have, since Margaret Thatcher came to power in 1979, been one of the main vehicles for the transformation of public power in Great Britain. Confronted with the economic slowdown and recession, the conservatives of the day made a case against state-centred corporatist regulation. Ideologically hostile to the whole Keynesian legacy and the welfare state, 'Thatcherism' brought in a clearly neo-liberal set of policies from 1979 to 1990. It eliminated exchange control, lifted price and wage controls, and then deregulated the capital market starting from 1986 by making drastic cuts in social expenditures, privatising whole sectors of the economy and public utilities and directly confronting the trade unions (the British miners' strike, put down in 1985, comes to mind). The free-market formula itself was by no means new or innovative. It could be found laid out in very similar terms during the same period in the United States under the Reagan administration, and again in New Zealand, which became, starting from 1984, a 'laboratory for testing free-market principles'. One of the particularities of Great Britain lies in the 'treatment' the government had in store for the cities. 'Thatcherism' was not just limited to a series of macro-economic measures guided by the principles of von Hayek. A territorial – urban, to be precise – foundation also existed. In fact, in the mid-1980s, the metropolitan institutions of British cities were disbanded by the Conservative government on the grounds that they were inefficient and contemptibly bureaucratic. New national programs aimed at 'helping' the cities restructure were designed and implemented. The referent used by the Conservative government changed: the economic crisis was not only sector-based; there was an urban crisis as well. To date, few other countries have so firmly rooted the solution to recession and the transformation of the state in urban policies. In Great Britain these action plans, metropolitan institutions, and new modes of political regulation were all vehicles for solving the problem. Great Britain made the transition from the Welfare State to the 'Workfare State' (Jessop 1994) essentially by making the cities the central issue, turning urban policies into a political and ideological conflict zone between the Conservative and Labour parties. This dynamic is clearly examined in the chapter that Booth devotes to Sheffield, archetype of the single-industry British city, having experienced the pangs of economic restructuring from the 1980s onwards. In the space of 10 years, Sheffield lost 44,000 jobs in the iron and steel industry. Municipal politics, historically dominated by the Labour Party, could not help but be transformed. Booth points out the city council's various phases of adaptation to the new economic order and to the urban policies launched by the Conservative government. After a phase in which the Labour elite hardened their resolve, taking on a veritable wrestling match with the central government, in the end, they bent to its demands, particularly in terms of creating new structures of governance which included private actors and adopting those policies which resulted in a very distinct tendency towards adhocracy – generating an elitist and notability-based system of political functioning – and towards the multiplication of decision-making bodies.

Newman and Thornley also highlight this conversion of the local Labour elite in their chapter on the partnership that the new mayor of London, Ken Livingstone, has been developing, since he was elected, with the economic actors in that city. Leader of the Greater London Council in the 1980s and nicknamed 'Red Ken' because of his direct opposition to Margaret Thatcher's Conservative government, Livingstone represents, almost to the point of caricature, the Labour Party's ideological move towards the centre and its transformation into New Labour under the leadership of Tony Blair. While his personal relationship with Blair has for much of his mayoralty been highly antagonistic, Livingstone nonetheless shares the same vision of political modernity as the current British prime minister, one which goes hand in hand with partnership and concertation with civil society and, above all, with economic actors. In the case of Livingstone, Newman and Thornley emphasise that this approach can be partly explained by the strong dependence, in terms of budgetary resources, of London's new metropolitan institution – the Greater London Authority, headed by Livingstone – on the central government. While having the indisputable legitimacy that comes from having been elected by direct universal suffrage, the new mayor of London has almost no resources of his own. He has chosen to lessen his dependence on the central government and to 'govern' by mobilising private actors, even adopting the popularised neo-liberal ideology and declaring, as cited by Newman and Thornley, that at 'the heart of the Mayor's job is making sure that London's success as a city economy continues. This requires more than just taking account of business issues in making decisions. It means forging an effective and productive partnership with business'. Thus can be seen the extent to which the ideology of this former high-ranking member of the orthodox wing of the Labour Party has evolved.

In her chapter on Toronto, Boudreau also puts forward the organic link that continues to exist between metropolitan cities and their responsible authorities, in this case, the provincial government of Ontario. The choice of metropolitan government structures is the direct result of a unilateral decision taken by Mike Harris's Conservative government, which opted for a municipal merger to which many residents of the former central city were opposed. The case of Montreal, developed in the chapter by Latendresse, illustrates this same dependence of Canadian metropolitan cities generally on the provincial level of government, a dependence which Andrew does not hesitate to describe as a 'shame' (Andrew 2000, p.100). As it happens, this situation is not particular to Canada (Jouve and Lefèvre 2002a; Jouve and Lefèvre 2002b).

The transformation of metropolitan institutions in Toronto and Montreal was brought about with the use of much rhetoric about the 'entrepreneurial city'. As such, the ideology of the government parties that opted for these mergers were of little importance, unlike the 1970s (Keil 2000). In Ontario, it was the Conservatives who chose to merge the municipalities, while in Quebec, it was the social democrats of the *Parti Québécois*. It is more in the impact of the reform that differences continue to exist. In the case of Montreal, Latendresse stresses the importance of resistance from within the political sphere to the process of municipal merger as well as the various institutional configurations that this reform has brought about regarding the mediation between elected officials and civil society. No doubt due to the fact that municipal reform in Toronto is older, but also, and especially because it was part of a process, launched in the mid-1990s, to transform the government of Ontario,

Boudreau further analyses how the metropolitan reform in Toronto was rendered possible by making use of a territorial political culture, carried by the left-of-centre municipalities, making consultation and the mobilisation of civil society one of the main elements of their action plans. In Toronto, the transformation of the metropolitan institutional framework fit into a pattern of classic partisan opposition between various levels of government, but also constituted a vehicle used by the Conservative Party (in office in Ontario since 1995) for imposing the 'Common Sense Revolution' so cherished by Mike Harris, premier of Ontario from 1995 to 2002; in other words, for adopting a neo-liberal agenda. Thus, Boudreau demonstrates the striking ambiguity of this policy. On the one hand, it purports to be more open to consultation, particularly in the planning of and through a participative budget, and through the promotion of diversity. On the other hand, it has actually resulted in centralising the metropolitan political system that, beyond any form of consultation, does not hesitate to launch large-scale urban projects aimed at making the economic capital of Canada more competitive, yet without stopping to consider the consequences in terms of social polarisation.

The adaptation of metropolitan cities to the new order brought about by the most recent changes in capitalism – production network, the increasing importance of the financial dimension of the production system, etc. – arises again in the chapter written by Fontan, Klein and Lévesque on the Montreal experience of economic restructuring. Like other metropolitan cities in North America and Europe, the former economic capital of Canada was and continues to be deeply shaken by the wave of de-industrialisation that accompanied the recession, beginning in the 1960s. While it currently has quite a varied business portfolio, the Quebec metropolis nonetheless saw its traditional production base severely hit by the recession which here, as elsewhere, resulted in both soaring unemployment rates (particularly within the industrial sector) and a devaluation of certain industrial spaces in the urban fringe which have rightly been described as 'abandoned or marginalised districts'. This spiral of decline was essentially curbed by the vigorous mobilisation of actors from civil society, particularly organised community actors from within the Community Economic Development Corporations (CEDCs) and the unions, creating financing tools that made it possible to prevent the closure of firms that were having difficulties and to assist in their recovery. This model of economic development was alternative, not only in that it was not based on the two main classic actors –the state and private enterprise – but also because the frame of reference for action in this social or solidarity-based economy was different, with priority being given to the fight for jobs rather than to growth. Taking the examples of reconversion in the southwest districts of Montreal and the revitalisation of the Dominion plant due mainly to the intervention of the *Fédération des travailleurs et travailleuses du Québec* (FTQ, Quebec Federation of Labour) through its *Fonds de solidarité* (solidarity fund), Fontan, Klein and Lévesque see in the partnership that developed in Montreal, regarding industrial redeployment, an alternative development model. In this model, the organised actors of civil society, representing the interests of the workforce, succeeded in structuring the terms of trade, mobilising resources and laying the foundations of a 'plural economy', such that several different 'models' of economic development could co-exist within a single metropolis.

The chapters devoted to France touch on themes that, for the most part, bear the stamp of decentralisation. Promoted as a major theme of François Mitterrand's first seven-year term of office, decentralisation is still at the top of the political agenda among the political and administrative elite in both central and outlying regions. It is currently in a phase of revival with the significant change to the Constitution and the strengthening of the authority of the regions in intergovernmental relations. Launched in 1982 by the Minister of the Interior, Gaston Defferre, it had the triple objectives of rationalising government operations, making elected representatives more accountable and bringing political decision-making closer to citizens, decentralisation in France was marked by a series of very significant changes in the balance of power and the organisation of metropolitan governments. From a managerial point of view, decentralisation mainly resorted to a contract-style of operation, a legal formula that first appeared in the 1970s. The contract-based approach to public policy of all kinds (urban planning, the environment, economic development, municipal policy, public safety, culture, etc.) had the effect of shifting more responsibility to intermediate political and administrative decision-making levels, and is based on the principle that proximity, a key theme of the government led by Jean-Pierre Raffarin, makes it possible to go beyond institutional divisions, generates compromises and adjustments, and better responds to the specific needs of local spaces. In his chapter, Warin reviews these dimensions of modern public action by addressing the dynamics at play in proximity-based, management-by-contract, public policies. In doing so, he identifies several constraints that call into question the efficiency of the contract approach. First, is the difficulty of harmonising various action plans that differ in their objectives, financing, and management styles. Secondly, there is the delicate nature of the 'dialogue' that develops between public services, that ensures the public interest in France, and the voluntary sector. And, finally, there is the individualisation of the delivery of public services that has led to an unprecedented increase in user requests in terms of new rights-claims vis-à-vis public services, resulting in a differentiated application of public policy with all that this implies in terms of equal treatment of citizens before the law and in terms of equity.

The question of the direct participation of residents/citizens in public policy-making is also discussed in the respective chapters written by Chabanet, on the implementation of municipal policy in Vaulx-en-Velin, a 'sensitive' town in the outskirts of Lyon, and by Toussaint, Vareilles, Zepf and Zimmermann, which deals with the redevelopment of a public space in Villeurbanne, a municipality which is also situated in the Lyon suburbs. The challenges of consultation faced in these two examples are radically different. In Vaulx-en-Velin, which symbolises the crisis experienced by the former working-class suburbs of large French cities, it is the issue of integrating young immigrants into the political system that is at the heart of controversy and political dynamics and that is sometimes expressed in radical or even violent ways. Urban policy, introduced in the 1970s and 1980s, has increased in importance in the French political and administrative life, to the point of requiring the creation of a ministry in its own right, charged with a seemingly impossible mission. It is in charge of solving many problems, namely that of social and economic exclusion, of new forms of poverty that result in a process of disaffiliation, and of the integration of immigrant communities into a 'single and indivisible Republic' that does not recognise the existence of any form of distinctiveness that would call into question its homogeneity, even though French society accepts cultural diversity (Wieviorka 2003). Under this policy, there has been radical confrontation between the representatives of disadvantaged social groups living in these 'districts of exile' (Dubet and Lapeyronnie 1992), , and a political and administrative system in which local elected representatives play a central role on account of the implementation of decentralisation laws and the transfer of many areas of jurisdiction from the central government. Taking the example of a residents' association in one district of Vaulx-en-Velin, Chabanet analyses in detail the meeting and the clash between opposing forms of legitimacy and reference territories. After presenting the problem of immigration in France, Chabanet addresses in detail the way in which these relations have taken shape over more than ten years. The consultation involved in an urban development project in Villeurbanne was conducted in a clearly more civilised manner. The stakes associated with this project were admittedly not as high as those related to the municipal policies in Vaulx-en-Velin. There was, for instance, no urban crisis nor social exclusion involved here, but there was nonetheless an attempt to modernise municipal politics by turning more readily to consultation with residents when making choices regarding urban planning, particularly when dealing with public spaces. Through a clinical examination of the consultation process with residents, Toussaint, Vareilles, Zepf and Zimmermann highlight the balance of power and domination at the centre of relations between local elected representatives and their technical services on the one hand and 'residents/citizens' on the other. Can the institutionalised processes of public concertation that structure political exchanges between the political sphere and civil society within the context of municipal policies, or in an undertaking as mundane as redeveloping a public space, enable change in the local political order and call into question the central role of elected representatives in the decision-making process? This is the question that runs across these two chapters.

Houk addresses another important dimension of the implementation of decentralisation laws in France: the reconstruction of metropolitan leadership through the transformation of exchanges between local elected representatives. To this end, Houk analyses the development and implementation of the 'Paris-Marseille-Lyon' Law of December 1982, which transformed the internal political organisation of these three metropolitan cities by establishing a sub-municipal level of political management, the district councils. At first envisaged by the socialist government of the day as a tool aimed at weakening the leadership of Jacques Chirac, then mayor of Paris, this measure led to very distinct changes in the balance of relations between various local political leaders. Gradually, the regulation mechanisms between the city councils and the *arrondissement* councils changed, particularly in Paris and Lyon. Councils for the *arrondissements*, while endowed with a limitied budget and few technical and administrative resources of their own, have became major political stake-holders, structuring partisan and personal battles within the political establishment in the various metropolitan cities. How did this development come about? What role do concertation and locally based management play in the process by which a sub-municipal level of government comes to assert its authority? Houk addresses these various questions in detail.

The chapters of this book vary in terms of scope. The approaches chosen by the authors also vary, in large part reflecting the content of the differing political, economic and social contexts of the issues discussed. The integration of these perspectives into a single coherent examination constitutes one of the main challenges of any comparative undertaking, particularly when the comparison takes place within the framework of a collective effort. But this is also what makes this study significant on both a scientific and human level, since it involves an opening up of a new set of questions and moves to uncharted waters. To borrow a useful expression from Négrier, comparison in the social sciences rests on an 'ethic of movement', which results in 'accepting the transformation of the subject under investigation through the dynamic of comparison' (Négrier 2003, p.10). The conclusion of this book takes this position, attempting to restate a general set of issues while drawing some general lessons from its chapters.

References

Anderson, J. (1995) 'The exaggerated death of the nation-state'. In J. Anderson, C. Brook and A. Cochrane (ed), *A Global World?* Oxford, The Open University, pp.65-112.

Andrew, C. (2000) 'The Shame of (Ignoring) the Cities', *Journal of Canadian Studies*, Vol.35, No.4, pp.100-114.

Badie, B. (1995) *La fin des territoires*, Paris, Fayard.

Balme, R., J.-L. Marie and O. Rozenberg (2003) 'Les raisons de la confiance (et de la défiance) politique : intérêt, connaissance et conviction dans les formes du raisonnement politique', *Revue internationale de politique comparée*, Vol.10, No.3, pp.433-469.

Beck, U. (1992) *Risk society: towards a new modernity*, London, Sage Publications.

Castells, M. (2000) *The rise of the network society*, Oxford, Blackwell.

Castells, M. and P. Hall (1994) *Technopoles of the World*, London, Routledge.

Comor, J.-C. and O. Beyeler (2002) *Zéro politique*, Paris, Mille et une nuits.

Courtemanche, G. (2003) *La Seconde Révolution tranquille*, Montréal, Boréal.

Dubet, F. and D. Lapeyronnie (1992) *Les quartiers d'exil*, Paris, Seuil.

Gret, M. and Y. Sintomer (2002) *Porto Alegre. L'espoir d'une autre démocratie*, Paris, La Découverte.

Hooghe, L. and G. Marks (2001) *Multi-level governance and European integration*, Lanham, Rowman and Littlefield Publishers.

Jessop, B. (1994) 'The transition to post-fordism and the Schumpeterian Workfare State'. In B. Loader and R. Burrows (ed) *Towards a Post-Fordist Welfare State?*, London, Routledge, pp.13-38.

Jouve, B. and C. Lefèvre (ed) (2002a) *Local Power, Territory and Institutions in European Metropolitan Regions*, London, Frank Cass.

Jouve, B. and C. Lefèvre (ed) (2002b) *Métropoles ingouvernables*, Paris, Elsevier.

Keil, R. (2000) 'Governance restructuring in Los Angeles and Toronto: Amalgamation or secession?', *International Journal of Urban and Regional Research*, Vol.24, No.4, pp.758-781.

Kymlicka, W. (1998) *Finding our way: rethinking ethnocultural relations in Canada*, Oxford, Oxford University Press.

Lamoureux, H. (1999) *Les dérives de la démocratie*, Montréal, Vlb.

Le Galès, P. (1999) 'Le desserrement du verrou de l'État?', *Revue internationale de politique comparée*, Vol.6, No.3, pp.627-653.

Loughlin, J. (ed) (2001) *Subnational Democracy in the European Union. Challenges and Opportunities*, Oxford, Oxford University Press.

Négrier, E. (2003) *Changer d'échelle territoriale. Une analyse politique comparée*, Montpellier, Université de Montpellier I, Centre d'Etudes Politiques de l'Europe Latine.

Norris, P. (ed) (1999) *Critical Citizens. Global Support for Democratic Governance*, Oxford, Oxford University Press.

Pharr, S.J. and R.D. Putnam (ed) (2000) *Disaffected democracies: what's troubling the trilateral countries?*, Princeton, N.J., Princeton University Press.

Polanyi, K. (1944) *The great transformation*, New York, Farrar and Rinehart.

Skocpol, T. and M.P. Fiorina (1999) *Civic engagement in American democracy*, Washington, D.C., Brookings Institution Press.

Wieviorka, M. (2003) 'L'idée de nation et le débat public', In R. Canet and J. Duchastel (ed) *La nation en débat. Entre modernité et postmodernité*, Montréal, Athéna, pp.65-76.

Chapter 2

The Transformation of Citizenship and Democracy at Local and Global Levels

Jules Duchastel and Raphaël Canet
Translation by Stuart-Anthony Stilitz

The origin of the word 'democracy' is *demos cratos*, a Greek term that means the 'power of the people'. In principle, modern societies are democratic and based on the principle of citizenship. Modern political institutions are democratic inasmuch as they embody compromises among social groups with divergent interests and manage conflict peacefully. The form of democracy deployed in each society varies according to the prevailing, context-specific, institutional structures.

In modern societies, democracy is usually State-centred and legitimate public authority is a monopoly of the State, 'a seat of a power that is incorporeal, though it purveys the power of the individuals on whose behalf it governs' (Burdeau 1970, p.53). Here, the indivisible and inalienable character of State sovereignty, which in its Rousseauist formulation embodies the 'common will', is the *sine qua non* for achieving the ideal at the heart of modern politics, namely the political equality of all citizens forming a nation in a particular territory.

With time, citizenship legitimised State power and provided the political impetus needed to create a social bond. With the advent of modernity and the idea of democracy, it became the preferred form of mediation between State and society (Canet and Duchastel 2003; Schnapper 1994). In a democratic system, the State has a legal monopoly on the use of public violence because a particular political community, the nation, legitimises this monopoly. This is why it is often said that the State is the legal embodiment of a sovereign nation. Ultimately, the regulatory power of the State depends on its territorially defined authority. Stated differently, the State obtains its authority from a community of citizens living in a particular territory.

While this theoretical construction of political organisation in modern societies is ingenious, it does not stand up to reality. As Rosanvallon points out, there is a 'persistent contradiction between political and sociological approaches to the principle of democracy. In political approaches, democracy establishes the power of a collective subject; however, sociological scrutiny weakens the collective subject and reduces its visibility' (Rosanvallon 2003, p.25). This paper will not deal extensively with critiques of formal democracy or the abstract universalism of Enlightenment philosophers, since they are already well known (Thuot 1998; Rosanvallon 1998). Suffice to say they remind us that the concept of citizen equality at the root of modern democracy remains an ideal and that there is a permanent hiatus between the modern philosophical-political vision of the world on the one hand and observable sociological phenomena on the other hand. This hiatus sustains the thesis that there are recurrent crises in citizenship, democracy, societal institutions and the model of integration. These crises also suggest the possibility of regenerating democracy at different levels and in diverse forms.

We hypothesize that transformations in forms of democracy are linked to changes in territorial status, political regulation and citizenship. First, we demonstrate that globalisation leads to deterritorialisation, thereby increasing the number and forms of democratic practice. This process erodes sovereignty and increasingly bypasses the State-centred model. We then demonstrate that political regulation fluctuates between two models, the government model and the governance model. This fluctuation leads to a reconceptualisation of democratic practice itself. Lastly, we probe two trends, the extension and intension of citizenship,[1] that give rise to its so-called 'incorporation';[2] the trajectory here is from the universal to the particular. These trends have had a decisive influence on forms of democratic participation. To conclude, we discuss the forms of democracy that are likely to emerge from these various transformations.

Territorial Transformations

The classic model of the territorialised nation-state allows for various approaches and forms of political institutionalisation. However, before discussing these approaches we need to outline the main features of the classic model. Territory is the third pole in a triad of poles we have devised to portray the modern State. It constitutes the living space of the nation identifying the physical limits of the power and authority wielded by all national institutions. Territory is an instrument of social enclosure, encompassing all dimensions of a nation's political and symbolic life.

A nation-state's territory has two facets: (i) on the one hand it is a unified entity, the basic political unit in a system of international relations based on the classic Westphalian model; (ii) on the other hand it is diverse, divided internally into political and administrative sub-units (Aron 1962). The politicisation of a nation-state's territory has both an external component (such as border disputes) and an internal component (such as jurisdictional disputes). A territory imposes limits on citizen membership and on the exercise of democracy. It demarcates the spheres in which States establish their legitimate power and bureaucracies apply legal standards.

National territory also has a fundamental symbolic function often found in political discourse. For example, States sometimes use territorial myths, including those involving ancestry, the natural environment, or the acquisition of new frontiers.

---

[1] Extension of citizenship means extending its application to various social categories (it becomes progressively widespread). Intension of citizenship means broadening the scope of citizenship (it becomes increasingly comprehensive).

[2] We introduced the concept of the 'incorporation' of citizenship to describe a specific way of acquiring rights. Here, citizens acquire rights not as individuals but (i) as persons belonging to a particular category of rights holders, or (ii) as collective actors (Duchastel, 2002; Bourque, Duchastel and Pineault, 1999).

These myths can cultivate an idealised image of a nation's geography, defend or challenge its internal jurisdictional arrangements and defend its sovereignty against external challenges. Thus, the concept of national territory is closely linked to every principle of legitimacy underlying the political institutions of modern society. Consequently, definitions and formulations of national territory tend to be highly discursive. Local territories or 'lands' compete, sometimes symbolically, with 'national territory' for authority in a community. When this occurs, discourse on national territory becomes even more ideological. A good example would be the invocation of 'the Rockies' as a Canadian symbol of national unity 'from sea to sea' to oppose the territorial separatism espoused by certain Québec sovereignists. Spheres of alternative identification can develop at both the sub-national level (provincial, regional or city) and the supranational level (regional, continental or global).

There are several ways of governing a territory, and each type of political system has its own approach. Jacobin France, which was state-centred, is usually invoked as the ideal type of nation-state. It exemplifies two important attributes of the State-centred approach. First, it designates a type of system that gives the State broad autonomy in governing society. Second, it designates an extremely hierarchical, centralised form of power that projects a homogeneous vision of national territory. This vision stems from the indivisibility of the French Republic, which theoretically guarantees all citizens equality under the law. (In reality, however, France's territorial structure – from the 1982 and 1983 laws on decentralisation to the recent constitutional amendment on the decentralised structure of the Republic (March 2003) – is much less monolithic than its idealised conception would suggest). By contrast, the federal models employed in the United States and Canada give a greater role to local communities and to the sharing of sovereignty among different levels of government (Théret 2003).

Thus, while democracy requires particular types of institutions, there is always a dialectic between the local and the global that culminates in a specific equilibrium among the different levels of power (Duchastel 2003a). Models of democratic representation and administration range from centralised to decentralised, and contain elements of both abstract universalism and concrete relativism. Local communities vary greatly in their degree of autonomy and in their ability to give voice to local democracy.

Globalisation has profoundly changed modern territoriality. In fact, the territorial sovereignty of nation-states has been eroding continuously for more than thirty years. On the other hand, globalisation may now be running out of steam, since the major economic powers, particularly the United States, seem to be de-emphasising globalisation and are considering new ways of exercising national sovereignty (Laïdi 2003). Underlying this erosion has been an increase everywhere in the five dimensions of global cultural flow identified by Appadurai: finance, population, media, technology, and ideology (Appadurai 2001). These trends in globalisation put to rest any lingering claims that borders are able to stem these flows and that the border concept is inviolable. The greater penetrability of territory strengthens the growing authority of the new actors (civil society associations and private sector organisations) in the reconstituted world system. The leading new actors are the deterritorialised economic powers (the transnational corporations) and the major international organisations whose expertise has tended to subsume State policy. In sum, the idea that territory is the sole criterion for delimiting national sovereignty is no longer valid. Henceforth, States may be said to belong to broad territorial groupings within which political authority is shared.

There is another reason why approaches based on national territory are obsolete. A region of one sovereign State can co-operate politically and economically with a region of a neighbouring State. This type of co-operation can even lead to the dismemberment of a nation's territories and to the formation of lateral interregional arrangements. Co-operation between regions that exist side by side is often based on trade interests or cultural affinities. Decentralised and proximity-based co-operation has already emerged in Francophone regions of the Organisation Internationale de la Francophonie.[3] It also exists in certain regions of Europe and elsewhere in the world.[4] The North American free trade negotiations, which began in the 1980s, fundamentally changed the nature of territorial relations in Canada. There was a realignment of commercial flows from the east-west axis (inter-provincial) to the north-south axis (involving Canada, United States and Mexico) (Boismenu and Graefe 2003). In sum, when a State's territorial sub-units co-operate to partially elude State authority and establish external relations, the territorial sovereignty of the nation-State is eroded.

Lastly, we need to examine two other trends in contemporary States: judicialisation and decentralisation. These trends are combating the widespread institutional imbalance that stems from the concentration of decision-making power in the executive branch of government and its technocratic structures. Social critics maintain that the concentration of power is responsible for the democratic deficit in the workings of government and the gulf that separates political elites from the electorate. While judicialisation and decentralisation may be viewed as alternative approaches to the concentration of power, they are also contributing to the fragmentation of the political community.

A posteriori judicial control over executive decisions offsets the marginalisation of deliberative assemblies in the political decision-making process. The purpose of this control is to ensure that public policies are aligned with the primary principles guiding a nation's collective life. These principles are promulgated in a nation's most authoritative texts and are the supreme expression of its entrenched legal order, constitution and charter of rights. With judicial control, tribunals replace assemblies as forums for resolving social disputes. However, in Canada judicialisation has an additional effect – it fragments the political community (Mandel 1996).

---

[3] These forms of co-operation constituted the focus of the *Premières rencontres internationales des régions francophones* ('First International Meeting of Francophone Regions'), organised by the Rhône-Alpes region, 2-5 October, 2002 at Charbonnières-les-Bains, France.

[4] In its June 2002 study, which discussed the numerous co-operation programs initiated by the Francophone Regions, the Centre Régional d'Études et de Sondages (CRES) identified nearly one hundred co-operation agreements implemented exclusively in the French-speaking world. For example, the Rhône-Alpes region, which is the most active in this regard, has twelve partnership projects with other Francophone regions, notably the province of Québec (Canada), the province of Khammouane (Laos), the province of Ho Chi Minh City (Vietnam), the City of Beyrouth (Lebanon), the region of Timbuktu (Mali), the region of Rabat Salé (Morocco), the governorate of Monastir (Tunisia), and the Voivodie of Malopolska (Poland) (CRES, 2002).

Fragmentation is the upshot of judicialisation, which is based less on the pursuit of collective interest (obtained through collective discussion among elected officials of a territory recognised as national) than on protecting particular rights. These rights belong to specific groups of rights holders (recognised by various constitutional texts and charters of rights) for whom belonging to a particular territory is not particularly relevant to the way they define their identity (Bourque and Duchastel 1996).

The democratic deficit can be redressed by decentralising political decision making. The guiding principle of decentralisation is subsidiarity, which avoids the shortcomings of judicialised social relationships by instead advocating the establishment of deliberative assemblies at the local level. In unitary States such as France, administrative deconcentration[5] and tutelage (to ensure that State power continues and is visible) exist alongside political decentralisation. Decentralisation allocates different types of powers to different levels of territorial jurisdiction; consequently, the relative influence of these jurisdictions varies. The increase in the number of spheres in which partial power is exercised (either in unitary States or composite States) reveals that there are numerous ways of apportioning sovereignty. Thus, as with judicialisation, territorially based, contractual division of powers is always accompanied by fragmentation of the political community; this holds true for federations and confederations, as well as for decentralised States.

Government and Governance

Transformations in territoriality have a direct impact on the exercise of sovereignty and on forms of political regulation (Badie 1995; McGrew 1997; Badie 1999). Three trends are re-shaping territoriality: supranationalisation from above; lateral arrangements between regions; and local fragmentation from below. Territorial reorganisation, whether it be internally directed or externally directed, raises the question of how to share sovereignty and the institutions of political rule, whose number is likely to increase. The concept of governance arises in social discourse as well as in scholarly discourse; it sometimes replaces the concept of government since it encompasses forms of institutional political rule linked to new conceptions of territory. We will discuss (i) the ideal-type of democratic government that is still influential among the political institutions of established nation-states, and (ii) techno-legal governance – a model that is gaining currency at both the supranational and local

---

[5] The two continuums, concentration-deconcentration and centralisation-decentralisation, facilitate the identification of different institutional arrangements in unitary States (France is a good example of the way this works). The concentration-deconcentration continuum refers to the propensity of the central State to assign, throughout its territory, agents whom it controls directly through the administrative chain of command. However, the State gives these agents decision-making power within a limited jurisdiction (by delegation). The aim in deconcentrating central power is simply to improve the efficiency of the government and of the administrative apparatus; it attaches no importance to acknowledging local characteristics. By contrast, the centralisation-decentralisation axis refers to the propensity of unitary States to acknowledge the existence of local issues for which affected populations must take direct responsibility, particularly through elected bodies at the local level (Debbasch et al. 1983).

levels. We neither claim that the second model has completely replaced the first, nor that national sovereignty has totally outlived its usefulness. We assume, rather, that sovereignty is now shared and that nations are currently in a transitional phase in which these two models of institutional political rule exist side by side.

At issue in the transformation of state-centred territoriality is the extent to which sovereignty can be shared. Political philosophy holds that modern sovereignty must be absolute (Duplessis 2003). In fact, it *is* absolute, but only because it is based on the prevailing system of nation-States and international relations. We make a distinction between (a) unitary States, France being the prime example (or at least it was until France adopted a law in 2003 stating that henceforth it was officially a decentralised state), and (b) composite States (especially of the federal type), of which the United States is the prime example. The composite or federal model is based on a division of powers that gives each level of government complete sovereignty in its respective jurisdiction. The concept of shared sovereignty is useful in analysing new forms of regulation that have arisen in the contemporary, 'globalised'[6] period. The distinguishing feature of the federal model is the fact that 'shared sovereignty' is subsumed under a national constitution that transcends its constituent parts. The model applies to a territorial framework found in a particular type of nation-State, in this case, the federal State (Canet and Pech 2003).

In modern societies, the concepts of sovereignty and political regulation are closely related. Societies with a truly modern form of sovereignty have democratic legitimacy and the ability to create appropriate institutions for themselves. The form of democratic government that a society selects will correspond to its modes of institutionalisation, which involve separating and differentiating the institutions of its political, economic and cultural spheres. Modern societies give precedence to the political sphere. Thus, a sovereign society's capacity for self-institutionalisation finds its supreme expression in representative government 'of the people, for the people and by the people'.

It is possible to portray democratic government as an institutional triangle whose three poles (the three spheres noted above) have a reciprocal relationship. The entire tri-polar configuration may be viewed as a social web linking citizens together. For example, the State (located in the political sphere) provides the legal framework for the economic sphere: it allows the capital/labour relationship to function effectively. In the socio-cultural sphere, the conceptual separation of private actors and public actors was a pre-condition for the emergence of the civic bond safeguarding the principle of democratic representativeness. As a final example, social issues emerge as a compromise between market forces (the economic sphere) and social forces (the socio-cultural sphere). Thus, an institutional triangle emerges with political, economic, and socio-cultural poles. The three vertices of the triangle represent, in turn, the civil link (between the political sphere and the economic sphere), the civic link (between the political sphere and the socio-cultural sphere), and the social link (between the economic sphere and the socio-cultural sphere). Although modern governments rely primarily on the political sphere (in which they create laws and the legal system), they

---

[6] Roland Robertson, 'Globalisation: Time-Space and Homogeneity-Heterogeneity', in Mike Featherstone, Scott Lash and Roland Robertson (eds.), *Global Modernities*, Sage, London, 1995, pp.25-44.

must also take into account the various complex relationships established among all three spheres.

The very complexity of this modern institutional architecture allows for a variety of concrete organisational forms. The specific conditions in each nation will shape its form of democratic representation and the allocation of sovereignty. What remains constant is the principle of national sovereignty, which bases its legitimacy on the existence of a community of citizens who are entitled to rights and freedoms but also have an obligation to participate, in some form or another, in the establishment of democratic government.

It would be overstating the case to claim that the governance model is the antithesis of the democratic government model. As we will see, governance is closely associated with the idea of participatory democracy, and is often vindicated by critiques of the more-or-less centralised institutions of representative democracy found in modern democratic societies. However, governance is basically a new form of regulation that was developed in conjunction with the democratisation of firms and later used by large international organisations to manage economic programs designed for emerging or developing nations (Gaudin 2002). Essentially, the governance model is based on the application of clear rules and standards. When the governance model is applied to nations, it requires a constitutional State. It also relies on effective administration and the principles of responsibility, accountability and transparency. Subsequent definitions of governance also embraced the concept of participatory democracy.

We will deal with the differences between the government model and the governance model in two stages. First, we will differentiate the respective structures of the two models in terms of the logic of their institutions and actors. Second, we will deal briefly with the ways in which their underlying democratic logic has been transformed. The discussion will also note differences in the way the governance model is applied at the global and local levels.

We have already demonstrated that the model of modern government is based on an institutional triangle linking together the economic, socio-cultural and political spheres, with the political sphere dominating. Our view that the political sphere is central to this model, both as a source of legitimacy and as a way of exercising power, in no way predetermines the degree of centralisation deployed by the model's institutions and political machinery. To say that the State is transcendent, in the sense that it can override society's other institutions, does not necessarily imply that State power must be centralised. As we have seen, a State can choose among numerous institutional forms, each of which is centralised to a different degree. Thus, it is not this particular characteristic that demarcates the government model from the governance model. Rather, it is the fact that the logic of the governance model subverts the logic of the government model. In other words, it overturns the institutional triangle in which politics dominates.

The triangle of governance restructures relationships among the institutional spheres. For example, the State is no longer the transcendent actor, but simply one of several actors (Jouve 2003). The wage relations that previously structured the economic sphere are now more or less abandoned; corporate actors become the State's partners of choice. At the transnational level, a new relationship is formed between, on the one hand, States and their international agencies and, on the other hand, economic associations (chambers of commerce, business associations). At the local level, local governments ally themselves primarily with economic interests to implement plans for local economic development. A third partner, civil society, comprising all other interests in the socio-cultural sphere, completes the triangle of governance,[7] though belatedly and with a limited role. At the global level, international organisations under pressure from social movements are increasingly inviting civil society organisations to participate in consultative bodies (forums and summits).[8] At the local level, governments are developing sectors in which groups with common concerns, such as user groups and interest groups, have the opportunity to consult with one another.[9]

Democratic government has greater legitimacy if civil society plays an important role in the economic and cultural spheres. In addition, democratic government expects responsible citizens to form a social bond. Governance, on the other hand, redefines civil society and the relationships among the three societal spheres. Instead of promoting a stronger social bond, it advocates confrontation among particular interests. In the logic of modern government, civil society complements the State. It represents citizens' economic, social, and cultural interests, which it seeks to validate in the political sphere. In the logic of governance, civil society is comprised of groups that are more or less disempowered. The governance model seeks to integrate a fragmented civil society into a consultative process in which economic, state and civil sector interests confront each other within a framework of asymmetrical power. It is

---

[7] Civil society has traditionally included an economic component. Since economic interests already play an important role in the triangle of governance, we view civil society here as comprised of voluntary associations excluded from the political and economic spheres (in their strictest sense).

[8] As stipulated in Article 71 of the United Nations Charter, the UN has always made provision, through its Economic and Social Council (ECOSOC), for a relationship with the non-governmental sphere (also called non-governmental organisations, (NGOs), civil society, the private sector, the unofficial sector etc.). From the outset, ECOSOC was obliged to grant consultative status to any NGO who requested it, and some NGOs obtained this status as early as 1946. It later adopted a broader definition of what constituted an NGO. The new definition was described in resolution 1296 (23 May 1968), and especially in resolution 1996/31 (25 July 1996). Its purpose was to facilitate the recognition, and especially the integration, into the UN system of various organisations and major summits. The term 'civil society' was first used in conjunction with the term 'NGO sector' at the Special Session of the General Assembly of the International Conference on Population and Development (Cairo + 5), held in New York from 30 June to 2 July 1999. The term applied to a broad spectrum of NGOs until the term 'private sector' came to describe yet another part of civil society; this new term emerged the Special Session of the General Assembly of the World Summit for Social Development and Beyond (Copenhagen + 5), which was held in June 2000 in New York. Today, 'private sector' NGOs constitute the second pole of the non-governmental sphere.

[9] During the 1990s, the Parti québécois developed a neo-corporatist strategy for Québec by setting up an organisational structure serving as an unofficial alternative to provincial and municipal deliberative assemblies. To this end, it organised participatory structures and related events including socio-economic summits, regional development centres, local development centres and local employment centres. The Liberal Party of Québec has since abolished these participatory structures, replacing them with locally elected structures.

no accident that the language of governance generally employs the term 'stakeholder' rather than 'citizen'.

We will return to the question of whether the governance model is democratic later. The governance model is needed when political institutions no longer meet the regulatory challenges associated with globalisation and societal fragmentation. For example, given the lack of relevant political institutions at the supranational level, there is a pressing need to introduce appropriate standards and rules. Appropriate policies are required at the local level to adapt to the vicissitudes of markets and economic development. The general trend toward judicialisation and greater technocracy has received broad support in governance proceedings, at least at the supranational level. Indeed, the idea of imposing standards and procedures on a group of economic and political actors is an integral part of governance, the role of the European Union being a case in point (Pech 2003). The function of governance at the local level involves enacting contingency standards better suited to the immediate environment. On both levels, however, the question of the legitimacy of governance has been raised belatedly and ineffectively. Consequently, the governance model has been unable to overcome the political and economic forces that stand in the way of increased democratic participation, which should, in principle, constitute the central feature of governance.

The Transformation of Citizenship

We will now deal with a third order of changes, having previously stated that these changes would give rise to new forms of democracy. We have just observed that the level at which citizenship is expressed has changed significantly, both at the supranational and sub-national levels. We also saw that the definition of citizenship has been influenced by the rise of governance. We will now: (i) provide an overview of the legal and political dimensions of citizenship; and (ii) describe the changes that have occurred in the extension and intension of citizenship and that have altered its original meaning, and (iii) demonstrate that these changes occur when society creates new political institutions relying increasingly on judicialisation and governance.

Citizenship is both a legal concept and a concept in political philosophy. As a legal concept, citizenship refers to the rights and duties that delimit the relationship between the State and individuals. By contrast, the expression 'rights and freedoms' aims to restrict the State to its primary functions, while the expression 'rights and claims' is associated with increased State interventionism. Citizenship also plays a critical role in the political legitimation of modern government and is based on the idea of forming a social bond. Ultimately, the legal and political dimensions of modern citizenship cannot be dissociated and, if we wish to understand the profound changes that have affected citizenship, we must consider them together. Indeed, the rights and duties associated with citizenship cannot really be understood except as they relate to a particular conception of society and political organisation (Duchastel 2003b).

Historically, citizenship has taken various forms. Thus, there have been distinct French, German and Anglo-Saxon models (Schnapper 2000). Models of citizenship have had a greater or lesser proclivity for abstraction, and have wavered between organicist and contractualist conceptions of the nation. It is possible to devise a continuum with, at one end, the principle of political legitimacy (abstract citizenship) and, at the other, community membership and civic participation (concrete citizenship). The idealised formulation at one end of the continuum is based on a universalistic vision of citizenship, while the formulation at the other end emphasises local characteristics and acknowledges the formative influences of culture and identity. All of these models have weaknesses: for some observers, models promoting individualism lead to anomie; for others, communalistic models are too restrictive.

Different conceptions of citizenship affect legal rights in different ways. It is possible to view citizenship as the upshot of changes in the rights and duties of social actors who partake in the institutionalisation of the three spheres of modern society discussed earlier. T.H. Marshall outlined the historical trajectory of these rights and duties, which have evolved in scope and meaning (1964). Marshall describes how rights and duties were gradually extended to all social categories, regardless of sex, age and social class. In other words, even though citizenship is a universal principle, in practice it has been granted to all members of society only gradually. Its evolving meaning has reflected changes in the rights of citizens. Marshall cites the evolution of citizenship in England to demonstrate the gradual broadening of its scope. In England, citizenship successively integrated various categories of rights: civil rights in the eighteenth century, political rights in the nineteenth century, and social rights in the twentieth century. While it would be a mistake to mechanically apply this particular evolutionary process to other social and historical contexts (Rosanvallon 1992; Schnapper 2000), the very existence of this process suggests that transformations in citizenship are inexorable, since citizenship is embedded in the idea of modernity. Modern citizenship is a force for emancipation inasmuch as it is universal: it tends to apply uniformly to all categories of citizens, extending its coverage to all concrete situations affecting people's rights and freedoms.

The transformations in citizenship over the last two centuries coincided with the introduction of new political, economic and socio-cultural institutions. Civil rights and freedoms served as a template for the development of the capitalist economy; political rights created the necessary conditions for the development of the democratic State; and social rights facilitated the development of the Welfare State. However, the introduction of social rights was a watershed in the history of universalistic citizen rights. Until then, these rights had dominated the declarations of rights associated with the economic revolution in England and the political revolutions in France and America. Although social rights gave rise to universal social policies, they were specific or particular rights, that is, they applied to the particular conditions of *groups* of individuals. Marshall employed the concept of 'incorporation' to designate the process by which groups became legally empowered to act as individuals. He uses the example of collective bargaining between unions and corporations (ibid, p.103). Incorporating rights means that the terms 'claimants' and 'holders of rights' can refer to a collective entity (groups) as well as to natural persons (individuals). Thus, extending citizens' rights changes them in two ways: universal rights give rise to particular rights, and individual rights are extended to collective rights. These changes create a paradox: the very universality of rights leads to their particularisation, and then to their collectivisation through the gradual recognition of all situations involving inequality or domination.

If we wish to understand the current dynamics of governance and participatory democracy, we need to examine the idea of incorporation in greater depth. The incorporation of citizenship does not weaken the universality of rights; natural persons (individuals) continue to be holders of legal and political rights and enjoy civil liberties. On the other hand, the incorporation of citizens' rights means that the nature of these rights, and of their holders, becomes more complex. In fact, the complexity surpasses anything Marshall was able to observe in his time. Citizens' rights have gone beyond social rights to include cultural rights, while civic citizenship has evolved into a form of citizenship based on identity (Kymlicka 2001). The growing number of social movements demanding that society acknowledge the existence of social inequalities has prompted demands for the 'right to be different' (Taylor 1997). Many charters of rights and freedoms have included rights based on identity and have recognised differences that are cultural, linguistic, biological, or associated with life conditions or choices. Rights holders have seen their status change in another way: since individuals can have several identities, beneficiaries can belong simultaneously to several claimant categories; this gives them access to an array of particular rights. This division of rights is not necessarily regressive. It could even be argued that it demonstrates the emancipatory power of universal human rights, since specifying particular rights brings many facets of social justice to public attention. That said, the fact remains that incorporation of rights inevitably results in more complex social relationships.

Before turning to the political impact of the incorporation of rights, it should be noted that the phenomenon affects not only private individuals (natural persons) belonging to more than one identity-oriented group, but also corporate entities: that is, incorporated companies that are recognised as fully fledged rights holders under civil law. While corporate entities start with the right to negotiate contracts, they eventually acquire all rights enjoyed by natural persons, except for a few, such as habeas corpus and the right to vote. The Canadian Charter of Rights and Freedoms provides a good illustration of the legal competence of companies to claim these rights, which include freedom of expression in advertising, and religious freedom applied to business practices. Chapter 11 of the North American Free Trade Agreement (NAFTA), which allows corporations to bring legal action against States in order to defend their right to free enterprise and free trade, is another example of how the behaviour of enterprises resembles that of citizens. It is not surprising to see companies asserting their 'citizenship', not only in terms of their rights but also in terms of the responsibilities associated with these rights.

As with transformations in territoriality, the incorporation of citizenship is associated with the judicialisation of social relationships and with the growth of governance.[10] The increasing judicialisation of social relationships refers to the major shift from democratic management to judicial decision-making. Canada's repatriation of its Constitution in 1982 and its entrenchment of the Canadian Charter of Rights and Freedoms in the country's fundamental law, provide an excellent illustration of this trend. The Charter refers to a very diverse set of rights; thus, using the Charter, parties who feel that their rights have been infringed upon are more likely to bring

---

[10] According to Cardinal and Andrew (2001) the second of these two movements could serve to counter the dangers of the first.

legal proceedings before the courts. In addition to general rights and freedoms, it includes linguistic and cultural rights, the right to non-discrimination in areas such as gender, disability and sexual orientation, and native peoples' rights. The critical element in this new constitutional structure is the right of various categories of justiciables (persons subject to the jurisdiction of a State's courts) to challenge laws adopted by parliaments based on their respective rights. Consequently, the balance of power has changed significantly, since there has been an unprecedented increase in the political influence of judges. To be sure, the increase in their influence has been inversely proportional to that of lawmakers. However, two crucial factors, aswell as the length of time that these factors have been in play, have affected the new equilibrium: the level of judicial activism on the part of the courts and the extent to which elected officials have assumed their legislative responsibilities. While the impact of these factors cannot be determined with exactitude, there can be no doubt that the logic of making claims based on specific rights has affected the functioning of political institutions.

The demands of social groups subject to injustices have resulted in increased recognition of their rights. This is why these groups attend public forums and demand greater participation in the political process. Thus, governance may be viewed as a form of political rule facilitating the articulation of these groups' interests. Therefore, the deliberative mechanisms associated with governance offer an alternative to judicialisation. However, the corporate logic underlying these groups' participation does not change, since those who participate in these mechanisms retain all the traits of incorporated citizenship discussed earlier. The organisations involved use defence-of-rights arguments to protect the particular interests of distinct groups. The citizen, defined in modern discourse as a political subject, is transformed into a moral subject.

By Way of Conclusion: The New Forms of Democracy

We have asked how democracy is affected when profound institutional changes affect a nation's sovereignty and legitimacy. The changes may be summarised as follows. The concept of national territory is being reformulated in numerous ways, raising the problem of how to allocate sovereignty. Two approaches to political rule are attempting to address the problem. One is based on the government model and proposes solutions such as decentralisation or the creation of supranational political unions; the other is based on the governance model and proposes solutions such as the creation of international governmental organisations or mechanisms for joint action at the regional and local levels. Citizenship, whose boundaries steadily expanded until the concept included a 'corporate' dimension, has come to embrace new forms of political participation. In so doing, it has transformed the principle of legitimacy underlying modern political institutions. The table below outlines six forms of democracy as they apply to different areas of political regulation. These are: centralised representative democracy; decentralised representative democracy; supranational democracy; corporate democracy; dissent democracy; and radical democracy. To explain these six forms, we have divided them into two models of political rule: the government model and the governance model. Together they account for all models of political organisation employed by nations. Each type of political rule presupposes specific institutional arrangements and methods of legitimisation.

Table 2.1 Forms of Democracy and Types of Regulation

Type of regulation		Forms of Democracy	
Government Model	National framework	Centralised representative democracy Abstract universalism *Jacobin Republic*	Decentralised representative democracy Concrete relativism *Federal and confederative models*
	Supranational framework	Supranational democracy Subsidiarity *European model*	
Governance Model	Techno-Legal regulation	Corporate democracy Participation *Summits and local consultation bodies*	
	Governmentality	Dissent democracy Resistance *Forums and counter-summits*	
		Radical democracy Indeterminacy *Spontaneous action*	

The essential distinction between these two models is the way they legitimise political action. In the government model, political legitimacy is based on the principle of representation, giving the decisions taken by the elected representatives of the political community (the source of sovereignty) their democratic character. The principle of representation also justifies State authority, either in its national or supranational form. In the governance model, political legitimacy is based on the principle of participation, which gives decisions their democratic character. In this model, designated or self-proclaimed representatives of various otherwise expert or relevant stakeholders take decisions. This principle rejects the primacy of the State, preferring instead either techno-legal regulation or new forms of action designed to counter the State's domination.

We will present each model of democracy in turn and discuss how they employ various features of local democracy.

Centralised Representative Democracy

Centralised representative democracy, a 'government' model of regulation, is the classic form of regulation associated with the modern State. The way it elects its representatives, who are given a mandate to raise issues of common concern, is predicated on the formal equality of all citizen-voters and the existence of a homogeneous, continuous territory subject to the sovereign and absolute power of the State. Decisions taken by central parliaments are supposed to reflect the general interest, and to apply uniformly over the entire national territory. Proponents of centralised representative democracy maintain that it is the consummate model since officials elected by the entire electorate make decisions. The model is based on an abstract universalism, and a central authority usually located in the political and administrative capital of the country makes all decisions; it therefore attaches little importance to local democracy. To ensure that decisions taken in the centre are fully implemented over the entire national territory, centralised forms of political rule often use deconcentrated forms of administration. France's Jacobin Republic is a prime example of centralised representative democracy. Under Law of 28 Pluviôse Year VIII (1800), the French Revolution made provision for the appointment of prefects, 'emperors with small feet' (i.e. officials with limited territorial authority), who were entrusted to serve as the local representatives of the central government. The decentralisation reforms currently underway in France are yet another indication that centralised representative democracy is becoming obsolete in the country.

Decentralised Representative Democracy

Decentralised representative democracy, the second 'government' model of regulation, allows for a broader range of political initiatives. In this form of democracy, sovereignty is not the exclusive prerogative of an overarching central State, as in centralised representative democracy. Rather, it is shared among several territorially defined entities, each of which is endowed with a legal personality. Thus, decentralised representative democracy is based on a concrete relativism that opens the path toward recognition of specifically local concerns. These concerns must be controlled by legitimately constituted regional or local governments embodied in assemblies of elected representatives, who deal with particular issues in their recognised areas of competence. By assigning different areas of competence to territorially defined entities, and then hierarchically ordering these entities, we can determine how they will share sovereignty; this also allows us to account for the equilibrium to be established among them. Decentralised unitary States, such as Italy, Spain and the United Kingdom (including Wales and Scotland), federations and confederations, such as Canada, the United States, Switzerland and Germany, are all models of shared sovereignty providing more or less extensive powers to the territorial entities within their respective domains.

Supranational Democracy

Supranational democracy goes beyond the national framework employed by the two forms of democracy discussed above. It aims to politically integrate nation-States located in the same supra-State regional grouping, while preserving the sovereignty of each member State. To this end, a complex institutional arrangement is contrived, combining various decision-making bodies located on several levels. On each body, representatives are elected on the basis of a direct or indirect vote. Supranational democracy is based on the principle of subsidiarity, which allows for non-hierarchical co-operation among contractually linked entities (national, supranational and infranational). Subsidiarity promotes local democracy, since it assumes decisions will be taken at the level most affected by the policies. This type of local democracy is dynamic and evolutionary; unlike decentralised representative democracy, which establishes the jurisdiction of each territorial entity *a priori*, supranational democracy determines the various 'levels of sovereignty' on an *ad hoc* and case-by-case basis. The creation of the European Union may be seen as an attempt to introduce a political order within an expanded regional government; it is therefore a good example of supranational democracy. Of course, at the time of writing, the European constitution is still being worked out and we hesitate to speculate on which institutional forms it will ultimately adopt. That said, Europe still views local democracy through the template of political modernity, though in a more complex way, since it completely reformulates the interrelationship of local, national and regional territories, at least as far as internal European matters are concerned.

Corporate Democracy

The fourth form of democracy is corporate democracy. It is a governance model embodying a form of regulation that bypasses legitimately constituted authorities organised on a territorial basis. Instead, it recognises non-hierarchical stakeholders whose sole reason for participating in collective negotiation is a manifest and recognised interest in the object of the negotiations. This form of democracy is based on participation by all parties interested in the idea of negotiation, and disregards the imperatives of democratic legitimacy and representation conveyed, at least in theory, by modern political institutions. It therefore abandons universal citizen representation, basing its actions instead on governance, which emerges as a techno-legal type of management bringing together actors who form groups to defend particular interests. To indicate the current sweep of this type of regulation we need only point to the various thematic summits and consultative bodies with a tripartite structure (comprising States and international organisations, the private sector and civil society) that are flourishing at every negotiating level (local, national, regional and global). This is a disparate collection of groups with varying degrees of structure.

Corporate democracy may be understood in two ways. On the one hand, it may be viewed as an outgrowth of the gradual trend toward enfranchisement (that is, the trend toward extending citizen rights). This trend has allowed 'incorporated' groups of highly mobilised citizens to effectively assert their rights through negotiations culminating in policy proposals that affect them. Thus, some view corporate democracy as a remedy for ineffective representative democracy and ossified political institutions. Others view corporate democracy as a purely formal exercise, bringing together a range of actors with extremely uneven power. In the latter view, democracy is a pretence: it hides behind the lofty principle of inviting all interested parties to the table in order to conceal an ideology favouring the interests of the strongest member organisations, usually from the private sector and the most developed nations.

Dissent Democracy

The next two forms of democracy, dissent democracy and radical democracy, draw their inspiration from the resistance strategy advocated by Michel Foucault. This strategy is based on Foucault's analysis of the mechanisms of knowledge and power, which he summarises in his concept of governmentality.[11] His analysis is relevant to the development of local democracy in that it advocates taking concrete action in all spheres of society in which knowledge and power are deployed instrumentally. Governmentality is completely opposed to any form of sovereignty over territory or over a territory's subjects. The ultimate objective of governmentality is to achieve the 'common good', but for Foucault this objective is tantamount to submission to the law.

Dissent democracy is associated with the governance model. As such, it rejects State sovereignty, since the latter is based on representation and is further weakened by an abstract formalism. Dissent democracy also denounces the State for excessive intervention in societal institutions and 'the lifeworld' and for using its knowledge to impose its power. It differs from corporate democracy in that it is critical of institutionalised tripartite partnerships giving rise to a techno-legal form of regulation. International organisations and summits have shown a growing affinity for these partnerships. In sum, dissent democracy is based on the principle of resistance. It is influential in the alternative forums and counter-summits that are currently exploring new forms of political action and power in contemporary society.

---

[11] For Foucault, governmentality comprises 'all instructions and procedures, analyses and deliberations, prognoses and tactics facilitating the exercise of a highly specific though extremely complex form of power targeting the population; its principal frame of reference is political economy, while its principal technical approach is the security structure'. Foucault maintained that these 'instructions... and tactics' have been introduced gradually to ensure that the power of government prevails over the power of sovereignty and discipline. This marked the transition from the 'Justice State' to the 'Government State' by way of the 'Administrative State' (Foucault, 1986).

Radical Democracy

The final form of democracy may be characterised as radical in that it pushes critiques of sovereignty, authority and power to their limit. Radical democracy is influenced by post-nationalism, post-colonialism and cultural diversity, and rejects all forms of determination and institutional constraint. It therefore opposes the State-centred approach to political organisation typical of modern societies, and abandons the idea of a transcendent political subject legitimising the sovereign authority of the State. Instead, the radical vision seeks to localise democracy spatially and culturally, and relies on spontaneous action. Thus, radical democracy is based on the principle of indeterminacy and the decentralisation of power. It rejects all forms of colonialist, hierarchical organisation, even the more benign forms that accept a degree of diversity (for example, Canada's multiculturalism policy and Charter of Rights). The discouraging note in this open form of democracy, in which everyone has the right to participate, is its predilection for anomie, taking root in the illusory belief that power can be repudiated or avoided.

References

Appadurai, A. (2001) *Après le colonialisme. Les conséquences culturelles de la globalisation.* Paris, Payot.

Aron, R. (1962) *Paix et guerre entre les nations.* Paris, Calmann-Lévy.

Badie, B. (1995) *La fin des territoires. Essai sur le désordre international et sur l'utilité sociale du respect.* Paris, Fayard.

Badie, B. (1999) *Un monde sans souveraineté. Les États entre ruse et responsabilité.* Paris, Fayard.

Boismenu, G. and Graefe, P. (2003) Le régime fédératif et la fragmentation des espaces, dans le contexte de la mondialisation. In J. Duchastel (ed) *Fédéralismes et mondialisation. L'avenir de la démocratie et de la citoyenneté*, Montréal, Athéna éditions, pp.215-238.

Bourque, G. and Duchastel, J. (1996) *L'identité fragmentée. Nation et citoyenneté dans les débats constitutionnels canadiens, 1941-1992.* Montréal, Fides.

Bourque, G., Duchastel, J. and Pineault, É. (1999) L'incorporation de la citoyenneté, *Sociologie et sociétés*, Vol.31, No.2, pp.41-64.

Burdeau, G. (1970) *L'État.* Paris, Seuil.

Canet R. and Pech, L. (2003) Fédération ou confédération? Les cas canadien et européen. In J. Duchastel (ed), *Fédéralismes et mondialisation. L'avenir de la démocratie et de la citoyenneté*, Montréal, Athéna Éditions, pp.95-114.

Canet, R. and Duchastel, J. (ed) (2003) *La nation en débat. Entre modernité et postmodernité.* Montréal, Athéna éditions.

Cardinal L. and Andrew, C. (ed) (2001) *La démocratie à l'épreuve de la gouvernance.* Ottawa, Presses de l'Université d'Ottawa.

Centre Régional d'Études et de Sondage (2002) *Régions et francophonie. Les coopérations des Régions françaises.* Grenoble, CRES, juin.

Debbasch, C. *et al.* (1983) *Droit constitutionnel et institutions politiques.* Paris, Économica.

Duchastel, J. (2002) Citoyenneté incorporée et nouvel espace des nations, *Revue d'études constitutionnelles*, Vol.7, No.1-2, pp.18-34.

Duchastel, J. (ed) (2003a) *Fédéralismes et mondialisation. L'avenir de la démocratie et de la citoyenneté.* Montréal, Athéna éditions.

Duchastel, J. (2003b) La citoyenneté dans les sociétés contemporaines. In J.-M. Larouche (ed) *Reconnaissance et citoyenneté*, Sainte-Foy, Presses de l'Université du Québec, pp.57-78.

Duplessis, I. (2003) La souveraineté politique. In J. Boulad-Ayoub and L. Bonneville (ed) *Souverainetés en crise*, Paris et Sainte-Foy, L'Harmattan and Les Presses de l'Université Laval, pp.17-27.

Foucault, M. (1986) La gouvernementalité, *Actes. Cahiers d'action juridique*, No.54, pp.6-15.

Gaudin, J.-P. (2002) *Pourquoi la gouvernance?*, Paris, Presses de Science Po.

Jouve, B. (2003) *La gouvernance urbaine en questions*, Paris, Elsevier.

Kymlicka, W. (2001) *La Citoyenneté multiculturelle: une théorie libérale du droit des minorités*. Paris et Montréal, La Découverte et Boréal.

Laïdi, Z. (2003) La crise de la gouvernance mondiale, *Le Devoir*, lundi 02 juin, p.A.7.

Mandel, M. (1996) *La Charte des droits et libertés et la judiciarisation du politique au Canada*. Montréal, Boréal.

Marshall, T.H. (1964) *Citizenship and Social Class and Other Essays*. Cambridge, Cambridge University Press.

McGrew, A. (1997) *The Transformation of Democracy? Globalisation and Territorial Democracy*. Cambridge, Polity Press.

Pech, L. (2003) La solution au déficit démocratique: une nouvelle gouvernance pour l'union européenne?, *Journal of European Integration*, Vol.25, pp.131-150.

Robertson, R. (1995) Globalisation: Time-Space and Homogeneity-Heterogeneity. In Mike Featherstone, Scott Lash and Roland Robertson (eds) *Global Modernities*, Sage, London, pp.25-44.

Rosanvallon, P. (1992) *Le sacre du citoyen*. Paris, Gallimard.

Rosanvallon, P. (1998) *Le peuple introuvable*. Paris, Gallimard.

Rosanvallon, P. (2003) *Pour une histoire conceptuelle du politique*. Paris, Seuil, 2003.

Schnapper, D. (1994) *La communauté des citoyens. Sur l'idée moderne de nation*. Paris, Gallimard.

Schnapper, D. (2000) *Qu'est-ce que la citoyenneté?* Paris, Gallimard.

Taylor, C. (1997) *Multiculturalisme. Différence et démocratie*. Paris, Flammarion.

Théret, B. (2003) Le fédéralisme, moteur ou régulateur de la mondialisation? In J. Duchastel (ed) *Fédéralismes et mondialisation. L'avenir de la démocratie et de la citoyenneté*. Montréal, Athéna éditions, pp.29-63.

Thuot, J.-F. (1998) *La fin de la représentation et les formes contemporaines de la démocratie*. Québec, Nota bene.

Chapter 3

Contemporary Cities and the Renewal of Local Democracy

Pierre Hamel

There has been a revival of interest in the theme of the city (Perry and Harding 2002; Andrew et al. 2002). This is undoubtedly due to the fact that growing numbers of people live in large urban areas. In 2001, according to Statistics Canada, 55 per cent of the country's population was living in one of Canada's 24 census metropolitan areas,[1] and in the country's four main urban regions, population levels had risen by 7.6 per cent from 1996 to 2001, compared with a growth rate of 0.5 per cent for Canada as a whole. Moreover, there have been increasing concerns about local institutions and their fragility in the new metropolitan context.

The contemporary cities in which we travel back and forth on a daily basis have retained many of the characteristics of modern cities, which the early sociologists associated with the fundamental issues of modernity, including its tensions and uncertainties. At the same time, contemporary cities are clearly very different in many ways. Indeed, while the modern city enabled individuals to experience a definitive break with the past – 'The modern city symbolised how we had uprooted ourselves from the past, from the traditions and miseries of our villages' (Lapeyronnie 1999, p.19; our translation) – the contemporary city is seen as an irrevocable reality, as a given. We clearly live in an urban civilisation, to paraphrase H. Lefebvre (1970), even when we live in semi-urban or rural areas.

In contemporary cities, the challenge lies not so much in building a capacity for self-affirmation or experimenting with a new identity that is torn between commitment and detachment than in increasing the capacity for self-expression in numerous areas – social, cultural, professional and political – where the main issue is that of recognition. This type of process can be linked to increased individualisation. U. Beck (1994) refers to the outdated modes of social integration associated with industrial society. Due to the rise of a 'reflexive modernisation,' he says, the modes of regulation associated with industrial society have become obsolete. This has resulted in the development of new forms of sociability, as individuals are obliged to assume greater responsibility – and a different type of commitment – in the crafting of their own biographies.

But the reality described by U. Beck (1994) merits closer examination. The individualisation that he refers to is undoubtedly not as universal as he implies, even in

---

[1] This term refers to urban centres with more than 100,000 inhabitants.

western societies, where communitarianism retains its appeal and is expressed in a number of concrete ways. We can think for example of the presence of ethnic enclaves or of the dissemination of communitarian values across urban areas or even on a broader scale, which in large part attenuates the individualisation of social relations.

Moreover, the transformation of social relations linked to individualisation does not eliminate the demands and requirements that we associate here with the issue of recognition. This transformation goes beyond a change in modes of sociability. Institutional structures and political approaches are clearly also affected. We are referring to the capacity of social actors to influence decision-making processes within institutions and to contribute to the definition of public-political space. How do we protect individual freedoms and minority rights, while also preserving the common good?

This question, which has been a topic of debate in political sociology in recent years, has also been raised in a concrete way in civil society by various types of social movements (Laclau and Mouffe 1985). What we are attempting to do now is to examine this question from a dual perspective, that is to say, on the one hand from the viewpoint of collective action, and on the other hand from that of the city, defined as a living environment and a place of integration. We have chosen to focus on cities because it is mainly in contemporary cities and metropolises that citizenship is being redefined, as a result of globalisation (Sassen 1999).

The chapter is divided into three parts. We start by pinpointing some of the characteristics of contemporary cities in order to better understand the challenges that they pose to the redefinition of citizenship. We then examine the social and political context in which contemporary cities are evolving in order to present the specific demands associated with the emergence of an urban citizenship. Finally, we look at the notion of urban citizenship and at some ways in which social movements have experimented with this notion. How is this urban citizenship affecting the legitimacy of social movements? And what does it mean for the future of collective action?

Contemporary Cities, Globalisation and Collective Action

When we look at the North American reality, we seem to be living less and less in cities in the traditional meaning of the term. Most people pursue their daily activities within very large metropolitan areas. In Canada, data from the most recent census, in the spring of 2001, showed that 79.4 per cent of the country's population was living within urban regions. One of the characteristics of these regions is the presence of metropolises whose configuration is based on a new definition of centrality – more dispersed and more discontinuous – which is associated with increased population mobility and greater territorial fragmentation than had been the case in the past (Biarez 2000).

These contemporary metropolises have superseded and are in many ways different from modern metropolises. Although the latter appeared at the end of the nineteenth century and in some cases dated back to the 1880s, some of their organisational principles were maintained right through to the 1970s. A number of researchers point to this decade as marking the decline of the Fordist model and its influence on spatial organisation (Filion 1995). Alongside the changes in production systems and lifestyles during this decade, new organisational principles and new approaches to urban planning began to emerge.

While modern metropolises assumed a hierarchical and centralised urban form that was defined from a dominant pole, contemporary metropolises follow a new mode of urban structuring. Today's metropolitan spaces are discontinuous, diverse and multipolar (Ascher 1998). Although the principle of networked spatial organisation associated with the modern metropolis is still prevalent, affecting both the location of activities in space and their social and cultural significance, this principle is now associated with a polynuclear urban system. And here we can think, for example, of edge cities (Garreau 1991), of urban fringes and their commercial nodes (Liebs 1985), or of continually expanding conurbations with multiple service areas.

However, the diverse forms now assumed by metropolitan spaces are giving rise to new processes of social and spatial segregation. Some American researchers consider urban sprawl and the concentration of poverty in urban cores as the two sides of a single metropolitan process (Dreier et al. 2001). In Europe, social exclusion processes are also associated with divisions and segregation, which are reflected not only in urban planning but also in the questioning by the middle classes of the social policies designed to reduce social and spatial inequalities that were developed in the context of the welfare state (Donzelot 1999).

The fragmentation of territorial systems that characterises contemporary metropolises is related to several factors, beginning with the restructuring of modes of production around a service economy. These changes can be linked to globalisation and the rise of the information-based economy (Castells 1996). However, we can also point to other factors on a social, cultural and political level (Ascher 1998). These include urban policies that favour suburban growth to the detriment of central cities; policies that often date back to the 1950s, yet continue to foster urban sprawl (Dreier et al. 2001).

Furthermore, the changes in the built environment in contemporary metropolises, which testify to a profound restructuring of the urban fabric, have had an impact on the renewal of approaches to urban management and urban governance. Some researchers (Lefèvre 1998) have even spoken of a 'renaissance' of metropolitan governments in Europe and North America that began in the 1990s. This was based on more flexible coordination structures between territorial units and on institutional choices in favour of negotiation and partnership, in the context of a dynamic process aimed at the development of coordination and intervention mechanisms that would be open to all the actors concerned. It is in this sense that one can speak of a new 'metropolitan governance' (OECD 2000). We can also speak of a new regionalism, including the Canadian version, which is different from American experiences in that it retains some of the features of the old, more directive type of regionalism (Sancton 2001).

On an economic and geographic level, the changes occurring in contemporary metropolises have been associated by a number of scholars (Knox 1997; Machimura 1998; Sassen 2000) with globalisation and its impact on spatial organisation and territorial planning. Although the impact of globalisation on cities varies widely according to local cultures, local resources, and local capacities to take advantage of prevailing economic trends; the local and the global are becoming more and more inextricably linked (Perry and Harding 2002). This has led to a restructuring of the urban hierarchy, that was in the past associated with industrialisation (Krätke 1992), and has given municipal administrations a new role in urban management and economic development in a context marked primarily by the international division of labour.

The consequences of these transformations can be interpreted in two different ways. On the one hand are the analyses that emphasise the negative consequences of globalisation, including its impacts on poor people, who have become more vulnerable due to the upheavals caused by the increased mobility of capital (Bauman 1998). Other researchers prefer to emphasise the positive aspects – the new opportunities – and especially the successes achieved by local actors able to overcome the structural changes and their negative effects. Harloe (2001) points out that our perception of urban environments has changed greatly since the 1970s and 1980s. Whereas cities (and large urban areas in particular) were seen by many as a problem rather than as a solution – as areas characterised by their less mobile populations and by their overall decline in a rapidly changing world – recent attitudes have completely changed. Cities are now often seen as dynamic economic centres and as hubs of creativity and innovation.

This kind of analysis is in line with the views of Jessop (1997; 2000), who feels that the new economic context, which fosters greater competition between cities, has prompted a more entrepreneurial approach to urban management. Cities are now required to take the initiative in developing economic strategies and are consequently having to improve their local management. They are being forced to re-examine their old management approaches, including their relations with both the market and with higher tiers of government, in order to succeed in the new context of heightened international competition.

In light of these changes, can we continue to maintain that 'the production of urban space is the result of collective action' (Fijalkow 2002, p.21, our translation)? What is the meaning of the notion of the city – as a socio-historical entity and as a political institution – in relation to the spreading urban system of the metropolitan area or city-region? In cities, is it still possible for local political actors to create new forms of solidarity that are appropriate for their inhabitants? Are we moving away from the idea of greater urban social justice, due to the tensions between the city centre and the periphery and the new socio-spatial inequalities engendered by urban restructuring related to economic globalisation?

These questions are not new. But they encourage us to look at issues in local democracy in today's metropolitan context from a new angle, in considering the issue of this democracy in relation to economic development. Like several other urban areas in north-eastern North America, Montreal – which has undergone numerous changes in recent years that can be linked to globalisation – can serve as an example to help us to better understand some current manifestations of the question of alliances and compromises made by local actors in a metropolitan area.

Like a number of other cities in the 'rustbelt,' Montreal is a former industrial centre. Many sectors of the old economy have had to be replaced with activities generally associated with the new economy. However, although Montreal's urban sprawl remains a concern for public managers, its effects have not been as devastating as those seen in many metropolitan areas in the United States, where they have been combined with the widespread deterioration of central cities (Dreier et al. 2001) and the resultant decline of the city in general (Beauregard 1993). But the stagnation of some traditional economic sectors, and the high level of poverty in some of Montreal's former industrial districts, in urban areas with large populations of immigrants and in some suburban areas, testify to problems in the economic and social integration of a number of social groups.

Montreal is a cosmopolitan city, to a greater degree than many other North American metropolises of similar size. Although in the past few years many immigrants have left Montreal, at various periods in the city's history, immigrants have played a crucial role in the dynamism of the urban area and have helped to redefine its cultural profile and its social geography. Apart from, but also alongside, the Francophone and Anglophone heritages that propelled the first phases of the city's development, the various waves of immigration have helped to shape what some have called a 'hybrid cosmopolitan identity' (Germain and Rose 2000, p.247). Since the 1980s, immigrants from many different parts of the world (Africa, Asia, the Caribbean, Latin America, Eastern Europe, etc.) have altered the social and cultural landscape of much of the urban area:

> 'the ethnic diversity of the new immigration has not simply added new ethnic neighbourhoods to the pre-existing mosaic. In fact, multiethnic neighbourhoods are now mushrooming in many parts of the Montreal region, some of them being new reception areas, others the product of major reconfiguration of the local ethno cultural profile' (Germain and Rose 2000 p.236).

This has resulted in a different cultural climate throughout the urban area from the atmosphere prevailing in the 1960s and 1970s; a climate fostered by a greater degree of ethnic and cultural diversity.

While this diversity reinforces the tendencies toward dispersal that characterise contemporary metropolises and creates the need for new mechanisms of social integration and social recognition, the question of social and political change is left open. In other words, Montreal's diversity encourages us to consider how public choices are made in the area of planning and development and the role that social and community actors, in the sense that we give to this term in Quebec, could play in this regard.

The Emergence of an Urban Citizenship

Urban areas are influenced in various ways by the economic, social and cultural processes that many scholars have associated with globalisation (Sassen 2000; Ascher 1998; Bassand 2001). Whereas in the past the traditional economic functions of cities were carried out primarily in local, regional and national contexts, urban areas are today acting more and more often as gateways to a global consumption network (Knox 1997). Cities also serve as strategic locations for social restructuring, not only because they are hubs for globalisation processes but also because they provide central spaces for managing globalisation, given the concentration of activities or functions that is required to cope with globalisation (Sassen 1999).

Globalisation affects local environments in very different and unpredictable ways (Clarke and Gaile 1998). In other words, globalisation has highly diverse local impacts. These impacts also lead to conflicts and political choices varying greatly according to local contexts, as Savitch and Kantor (2002) show in their comparative study of several European and North American cities. But this does not prevent the tendencies associated with globalisation from challenging the policies of central governments in terms of their capacity to define the local production conditions required by the trans-national flow of capital, as these researchers argue. In this sense, we can say that such changes raise challenges in the sphere of local government. The new realities are inducing local actors to call for a new form of power sharing between different levels of government and for changes in the functioning – if not the very basis – of regulatory mechanisms, so that local authorities are granted more room to manoeuvre and greater powers.

Clarke and Gaile (1998) speak of a new capacity for local action, of a 'new localism', that is prompting local actors to re-examine their position in the new urban hierarchy emerging from the tendencies toward globalisation, and to redefine their economic and political role. What is at stake here, and what local governments are trying to promote, is an improvement of the economic and social context in which their cities operate. To achieve this objective, they are ready to negotiate with a wide range of actors and to base their strategies on both innovation and cooperation. In this way, they are choosing to assume entrepreneurial functions, and to become true catalysts for development. Moreover, by taking strategic action, cities are also able to improve their position in the global hierarchy of cities.

But we cannot completely ignore the 'dark side' of globalisation which, as Clarke and Gaile (1998) stress, has also led to a new international division of labour, greater competition between cities and, often, new institutional arrangements of the type seen with the adoption of the North American Free Trade Agreement. Agreements of this kind are undermining national and local mechanisms to protect existing businesses and producers, to the advantage of more mobile investments and businesses. This has resulted in new types of inequalities, and in social and spatial polarisations (Bauman 1998; Clarke and Gaile 1998).

In this chapter we are not attempting to discuss all the ramifications of a complex situation that is rife with uncertainty, or to identify all the possible consequences for local actors. Our primary aim is instead to circumscribe the general effects of the new urban realities from a normative and political standpoint.

We can begin by emphasising that the nation-state, which served to define the bases of the political community and which can be associated with the 'first modernity,' in the sense given to this term by Beck (2000) – and which emerged in an era when society and the state were co-extensive, with the state exercising power and control over society in reference to a clearly defined territory – proved to be a true and powerful 'container of social processes' (Sassen 1999, p.134). It is this 'first modernity,' along with the values and compromises out of which it was defined, that was challenged by the rise of the powerful forces and processes of globalisation. From then on, the political rules changed, beginning with the mechanisms of control and regulation that nation-states had succeeded in establishing.

Globalisation is not an entirely new phenomenon, and it is not without uncertainty or ambiguity. In its recent manifestations, globalisation is profoundly transforming the values and conceptions of the social order that were associated with the 'first modernity.' This is what we find when we examine the economic, social and cultural exchanges that are occurring in large urban areas. Moreover, the basic premise according to which states and societies were created on separate, independent territories has suddenly become obsolete, or at least in some of its applications. Borders are becoming more permeable and less relevant, as daily practices have become more open and more affected by the transnational flows of goods, capital, cultures and individuals:

> 'Globality means that the unity of national state and national society comes unstuck; new relations of power and competition, conflict and intersection, take shape between, on the one hand, national states and actors, and on the other hand, transnational actors, identities, social spaces, situations and processes.' (Beck 2000, p.21).

Increased international flows across national borders are making it more difficult for national governments to exert control over their territories (Urry 1998). The very foundations of democracy, as they have been historically associated with the nation, are being challenged. As D. Schnapper points out:

> '[T]he democratic nation is being weakened because its sovereignty in the world of nations and political units is becoming increasingly limited; it is being weakened because its political objective of integrating populations is becoming irrelevant, as the national reality is gradually being transformed into a community of work, culture and redistribution of wealth' (1994, p.185; our translation).

It seems appropriate to take a closer look at the transformations affecting the political community, starting with one of its main foundations; citizenship.

But this weakening of the democratic nation that D. Schnapper refers to – and which we can associate with the various mechanisms that are transforming the formerly unquestioned might of the nation-state – is not an exclusive and definitive process. It is still too soon to announce the death of the nation-state. Nevertheless, we are seeing, more and more often, a variety of experiments centring on citizen participation and the creation of new types of institutional relations in the development and management of public policy (Berry 1999). And here we can point to the mobilisation of social movements that is occurring in the sphere of civil society and is being accompanied by the demand for more direct means of participation in public affairs than those advocated in technocratic management approaches.

These social actors are helping to shape a political agenda which includes ethical concerns that are moving away from those traditionally advanced by the liberal elite (Offe 1997). These actors are calling for a different kind of integration of a culturally diverse citizenship and of the multiple identities characteristic of the societies of late modernity. If globalisation is shaking the political foundations of the state, it is also encouraging the expression of a pluralism that is leading to a re-examination of the traditional definition of solidarity, which was primarily seen as the responsibility of the state.

Clearly, by promoting a politics of identity, the new social movements of the 1970s had already drawn our attention to the limitations of top-down modes of integration, which we can associate with an abstract definition of citizenship. In the wake of the women's movement, they also highlighted the idea that there was no democratisation of the public sphere without a democratisation of the private sphere. 'The democratisation of decisions and decision making at home' (Phillips 2000, p.445; our translation) was seen as a prerequisite to democratic equality. Participation in democratic life consequently necessitated a re-examination of the relations of power prevailing in the private sphere. Through individual behaviours or collective action, personal and public concerns were mutually reinforced. Civil society was now seen as the meeting place – as the space for the solidarity and compromise associated with social and political mobilisation and collective action.

The themes of self-actualisation and the recognition of cultural or sexual differences are profoundly transforming the political sphere and, at the same time, is encouraging us to re-examine the conception of citizenship as defined by T.H. Marshall, which is unable to take into account new forms of political subjectivity (Procacci 2001). Moreover, as the issue of citizenship is placed in the realm of civil society rather than in the realm of the state, we are moving away from overly abstract definitions of citizenship that prevent us from understanding the ongoing transformations in social relations as they are expressed in public-political space (Ku 2002). Due to the questioning of processes of regulation associated with the state, a questioning that globalisation is tending to accelerate, we are seeing an affirmation of identities in the political sphere according to their various forms of expression. At the same time, we find that the issues involved in democratisation are changing or, at least, that these issues are being defined in different ways due to an increase in social and cultural differentiation.

Hierarchical and centralised representations of the political sphere are gradually being replaced by a pluralistic viewpoint that encourages the expression of personal freedoms and subjectivity in a pragmatic context where the general interest is found in a negotiated compromise of the multiple interests involved. A passive notion of citizens' rights, conceived on the basis of the prerogatives of the state, is being replaced by an active vision of citizenship, which is defined on several levels (environmental, cultural, cosmopolitan or with reference to mobility) (Urry 1998). This leads us to consider the emergence of an urban citizenship (Roth 1998), which requires that we re-examine public participation mechanisms, in looking at the fragmentation of social actors in the sphere of civil society in a new way (Roth 2000). This notion of urban citizenship can help us to understand the issues involved in the social and political restructuring of contemporary metropolises. In that sense, we can hypothesise that the notion of urban citizenship may be useful in the current context of globalisation.

In the sense in which we are using it here, urban citizenship centres on the capacity of local milieu and local actors to change social relations with space and the city. It favours the use of individual and collective action to bring social and cultural concerns into public-political space, such as greater social justice, heritage preservation, or the promotion of local democracy (Clarke and Gaile 1998). From this point of view, urban citizenship is strongly influenced by the contexts in which it exists. How do we face the challenges of globalisation, which require more competitive milieu, while also combating the new types of inequalities associated with globalisation? How can the diversity and pluralism that characterise the social and cultural landscape of contemporary metropolises contribute to the definition of the 'right to the city,' so that it also recognises the 'right to differences' (Sandercock 1998)? What can we learn from mobilisation experiences or from various forms of citizen participation and collective action regarding the urban issues that characterise contemporary metropolises? And finally, what responsibility should public authorities assume for the future of these metropolises?

Local Democracy and the Question of Inequalities

In North America – and to a lesser degree in Europe – the urban development strategy, generally favoured by economic and political elites in contemporary metropolises, has focused on growth (Logan and Molotch 1987). But although there has been a very strong consensus for a number of years on the need for growth, some researchers note that this strategy is beginning to be questioned, even in cities such as Los Angeles where the coalition for growth has long dominated land-use policy (Purcell 2000). We can use this observation as a starting point, as we now turn to the types of conflicts that have arisen in contemporary cities and metropolises in recent years and at how these conflicts testify to the challenges faced by local democracy.

Since the 1960s, in all western countries, urban planning and urban affairs have engendered a number of controversies. Social mobilisations against projects promoted by developers or public authorities have led urban movements to formulate a social vision of the city that opposes the market-oriented view generally advocated by local elites. This has resulted in processes of modernisation of the public management of cities, as municipal administrations have agreed, to varying degrees, to make room for active citizen participation in urban affairs. But it is difficult to assess the impact that urban movements have had on the changes in urban policies. This is, in part due to the frequent merging of collective action with other contextual factors (Pickvance 1985).

Moreover, citizen participation is clearly not a panacea. For example, it does not guarantee that urban services will be provided with greater efficiency or equity (Wolman 1995). Even if it is not the catastrophe that some observers predict in terms of hampering decision-making processes, neither does it necessarily lead to lasting changes in the structures or mechanisms of power.

However, urban movements have certainly played an important role in the modernisation of urban planning and of municipal administration in general, in both their aims and their democratic functioning. At least, this is what we have seen in the case of Montreal (Hamel 2003); and similar processes have also been observed by researchers in other countries (Castells 1983; Fisher and Kling 1993; Mayer 2000).

In promoting social housing, heritage preservation, local development and local services that meet the needs of people in working-class neighbourhoods and various types of users in a number of cities in the world, urban movements have defended a dynamic conception of urban citizenship, that is, one that is open to social and cultural differences (Hamel et al. 2000).

At the same time, we have to recognise how dissimilar these urban movements are. They differ widely, not only in their attachment to diverse ideological currents, but also in their organisation and action strategies. Montreal's urban movements are a good example of this. Some of their mobilisations have been undertaken in the context of large coalitions, including labour union representatives – as, for example, in the struggle for tenants' rights during the 1960s – and others have been organised in the form typical of 'new social movements', as seen in some of the protests against the demolition of rental housing in the early 1970s, and more recently, in the summer of 2001, with the occupation of abandoned buildings by the homeless. Traditional issues, such as social assistance for the disadvantaged, are sometimes linked to new forms of poverty and growing factors in social exclusion, which some observers have associated with globalisation. At other times, some social actors have advocated direct confrontations with the symbolic holders of power, as seen in the protests against the Multilateral Agreement on Investments (MAI or, in French, AMI), the Millennium Round of trade negotiations, and, in particular the case of the civil disobedience advocated by the proponents of Opération SalAMI (Lemire 2000), whereas other social actors have chosen a more pragmatic approach. Some have attempted to change institutions from within. This is the kind of strategy favoured by community organisations involved in local economic development, as seen not only in Montreal but also in other cities in Canada and the United States.

A central concern of urban movements has been to challenge the relations of domination and power that have been exhibited – and reproduced – in urban planning and development. Urban movements have mainly attacked the 'commercialisation' of the city. They have also attempted to change the framework of local public management by trying to shift the focus from new modes of governance – although this path has also been explored through various experiences in participatory local service management and community action on local economic development – to a call for more public debate and deliberation. They have thus brought new topics and new concerns into the public sphere. Some well-known examples include the issues of tenants' rights, heritage protection, and women's safety at night. But urban movements have also led to the establishment of various formal mechanisms for citizen participation, public consultation, and partnership with the primary objective of broadening decision-making processes in regard to urban policy.

To what extent are such mechanisms – which are relatively recent in the case of Montreal in that they were only officially adopted in the late 1980s – opening the way for a real democratisation of urban management and urban planning, or are they merely a sophisticated smoke screen enabling elected officials to renew the status quo and to reproduce 'political and social relations of domination,' as Blondiaux and Sintomer (2002, p.33) have asked? Between functionalist perspectives, with their instrumental analyses of decision-making processes, and normative points of view developed in reference to the debate on the transformation of the democratic framework in complex societies, there is room for more nuance interpretations given the socio-political changes that have resulted from the setting up of these mechanisms in various contexts.

However, we must not forget that urban issues continue to be fuelled by social inequalities and indeed by the class relations that help to influence public policy, in spite of the generally accepted discourse on social exclusion and social cohesion, in the direction of dominant interests (Harloe 2001). From this point of view, the question of democracy and its new forms of expression in the urban arena does not eliminate concerns about social justice. Nevertheless, due to the many changes on the urban scene in recent years, which many observers associate with globalisation and metropolitanisation, we must now consider the problems of the city from a new angle.

In many countries, as Bassand (2001) has emphasised, the processes that we associate with metropolitanisation – urban sprawl, the accelerated movement of persons within urban areas, the integration of economic and social activities on a larger scale than in the past, the emergence of a discontinuous and more dispersed urban centrality, problems in terms of political representation – are triggering various types of crises. We are seeing crises in transportation, the environment, and in the location of regional public services and facilities. But there have also been repercussions in the political sphere, where a 'democratic deficit' is evident.

However, few countries, despite a veritable 'renaissance (in the 1990s) of institutional reforms in large cities' (Jouve and Lefèvre 1999, p.9; our translation), have opted for fully-fledged metropolitan governments. They have preferred to adopt hybrid forms of management and planning that can be seen as 'institutional compromises,' rather than to set up a new, integrated system of representation and decision-making. The principle of the network and of the coordination of activities from organisations that possess a certain degree of legitimacy in the political system is thus given precedence over the principle of centralised leadership in the creation of a new tier of government. This is what the proponents of the new regionalism in the United States are suggesting in calling upon the notion of governance.

The metropolitan question can be considered from various angles. From a socio-political viewpoint, we find that two perspectives converge, that is, issues of governance on the one hand and issues of citizenship on the other. In a schematic way, we can say that metropolitanisation has limited the traditional role of local administrations, which have had to coordinate their actions more than before with other nearby municipalities or to share their authority with sectoral management bodies acting on a regional scale. In this regard, we can speak of a new model of governance that in part escapes direct democratic control (Burns 2000). The question that then arises is whether citizen participation, through consultation mechanisms and public debate, is sufficient to offset the deficit in legitimacy that results from metropolitan governance. Are the same criticisms applicable here, from a democratic viewpoint, as those that were expressed in the past with regard to neo-corporatism?

This question leads to another, that is, the question of urban citizenship and its meaning in the metropolitan context. As we have already noted, urban citizenship is associated with an openness to multiple identities and to new forms of solidarity that must now be defined in light of the complexity of social and political relations in the age of cosmopolitanism. But despite its flexibility and its sensitivity to social and cultural differences, urban citizenship, as it has been defined and redefined by urban movements from the 1960s onwards, has in recent years been confronted with new social and political demands as a result of metropolitanisation. How do we develop alliances and create solidarity among populations that not only do not share the same social origin but also live out their daily lives in spaces – urban neighbourhoods or cities – that are often quite distant from one another? In other words, can urban citizenship also be expanded into a form of metropolitan citizenship?

This question has been raised by a team of researchers in Switzerland who conducted a survey of residents of four urban areas in that country (Kübler et al. 2002). The conclusions drawn by the researchers in their preliminary report are quite a nuance. It would seem that for residents in these urban areas – Lausanne, Lucerne, Zurich and Lugano – metropolitan space is not merely a functional space from an economic viewpoint. It is also becoming relevant to social practices. Residents are quoted as having a strong 'psychological attachment' to their metropolitan areas, although they are often unsure of the boundaries of the latter. In other words, metropolises are giving rise to the development of certain community bonds. They are helping to build a new collective identity on a territorial scale that is situated between that of the local and the national. On the other hand, these bonds are rarely defined in political terms, that is, in reference to objectives developed on the basis of common interests. In this regard, local concerns very often remain an important focus.

Moreover, the relations between issues of citizenship, legitimacy and service provision are still quite weak. The level and quality of services provided to the public and the modes of service provision adopted by public authorities do not have a significant influence on the legitimacy of local institutions. On the other hand, contrary to a hypothesis frequently advanced by urban studies researchers, the use of governance does not appear to represent a threat to democratic citizenship or to political legitimacy, although situations obviously vary depending on local contexts. Finally, we must also consider the effect of size. The four urban areas involved in the study cited above are much smaller than the North American metropolises mentioned previously.

Conclusion

In light of the observations outlined in this chapter, urban citizenship does not appear, at least in principle, to be incompatible with metropolitan citizenship. The fact that the city is assuming a more dispersed form than in the past – which was in any case already present in the 'network' city of the late nineteenth century – does not prevent the potential emergence of 'a sense of territorial belonging' (Fijalkow 2002, p.110; our translation). Nevertheless, the sense of attachment to metropolitan areas is often rather vague and does not rule out an attachment to local spaces, at the sub-regional, municipal, or neighbourhood level. Moreover, in many North American metropolises, political leaders responsible for metropolitan management are counting on the market as well as on social solidarity in promoting their territories (Fontan et al. 2003).

This has in recent years led to the public-private partnership that has emerged in urban management and local economic development. It also explains the growing number of governance experiences where public authorities have been willing to act as facilitators or participants rather than as leaders. Consequently, the tensions between public and private interests and the ways of managing these tensions are evolving in new directions. Social and political actors are finding it necessary to reconsider both their past alliances and their action strategies.

Urban movements, which since the 1960s have contributed in various ways to the democratisation of urban management and urban planning and have helped to transform the models of governance that have recently appeared in contemporary cities and metropolises, are now facing new challenges. Among these challenges is the need to take part in the definition of a metropolitan citizenship that will be able to link the issues involved in urban citizenship with a new social, cultural and territorial reality.

Although urban movements have in the past played a part in broadening and extending democratic mechanisms and processes on the local political scene, these gains have proved to be fragile, due to globalisation and the transformations that it is causing. Moreover, the advances made in the area of improved living conditions must now be placed in a new context. The city has changed, and new problems have arisen. Today, more and more often, these problems are being defined on a broader scale, on the level of large urban areas. This is the case, for example, with the new forms of urban poverty and the inequalities experienced by certain social groups, the integration of immigrants and cultural communities, protection of the environment, and the social, economic and political dynamism of metropolitan areas. In this regard, the contribution made by actors in civil society, especially social movements, is just as important as that made by elected officials and public administrators. And it is in large part in this new sphere of metropolitan governance that the future of collective action and of contemporary metropolises will be played out.

References

Andrew, C., K.A. Graham and S.D. Phillip (eds) (2002) *Urban Affairs Back on the Policy Agenda*. Montreal and Kingston, McGill-Queen's University Press.

Ascher, F. (1998) *La République contre la ville. Essai sur l'avenir de la France urbaine*. La Tour d'Aigues, Éditions de l'Aube.

Ascher, F. (2001) *Les nouveaux principes de l'urbanisme*. La Tour d'Aigues, Éditions de l'Aube.

Bassand, M. (2001) Métropoles et métropolisation. In M. Bassand, V. Kaufmann and D. Joye (eds) *Enjeux de la sociologie urbaine*, Lausanne, Presses polytechniques et universitaires romandes, pp.3-16.

Bauman, Z. (1998) *Globalization. The Human Consequences*. London, Polity Press.

Beauregard, R.A. (1993) *Voices of Decline. The Postwar Fate of US Cities*. Oxford and Cambridge, Blackwell.

Beck, U. (1994) The Debate of the 'Individualization Theory' in Today's Sociology in Germany, *Soziologie*, No.3, pp.191-200.

Beck, U. (2000) *What Is Globalization*. Cambridge, Polity Press.

Berry, J. M. (1999) *The New Liberalism. The Rising Power of Citizen Groups*. Washington, Brookings Institution Press.

Biarez, S. (2000) *Territoires et espaces politiques*. Grenoble, Presses Universitaires de Grenoble.

Blondiaux, L. and Y. Sintomer (2002) L'impératif délibératif, *Politix*, Vol.15, No.57, pp.17-35.

Burns, D. (2000) Can Local Democracy Survive Governance?, *Urban Studies*, Vol.37, No.5-6, pp.963-973.

Castells, M. (1983) *The City and the Grassroots*. Berkeley, University of California Press.

Castells, M. (1996) *The Rise of the Network Society*. Oxford, Blackwell.

Clarke, S.E. and G.L. Gaile (1998) *The Work of Cities*. Minneapolis, University of Minnesota Press.

Donzelot, J. (1999) Liens sociaux et formes urbaines. La nouvelle question urbaine. In T. Spector and J. Theys (eds) *Villes du XXIe siècle. Entre villes et métropoles: rupture ou continuité?*, Paris, Ministère de l'Équipement, du Transport et du Logement, pp.46-60.

Dreier, P., J.H. Mollenkopf and T. Swanstrom (2001) *Place Matters. Metropolitics for the Twenty-First Century*. Lawrence, Kansas, University Press of Kansas.

Eade, J. and C. Mele (2002) Understanding the City. In C. Mele and J. Eade (eds) *Understanding the City. Contemporary and Future Perspective*, Oxford, Blackwell, pp.3-23.

Fijalkow, Y. (2002) *Sociologie de la ville*. Paris, La Découverte.

Filion, P. (1995) Fordism, Post-Fordism and Urban Policy-Making: Urban Renewal in a Medium-Size Canadian City, *Canadian Journal of Urban Research*, Vol.4, No.1, pp.43-72.

Fisher, R. and J. Kling (eds) (1993) *Mobilizing the Community. Local Politics in the Era of the Global City*. Newbury Park, Sage.

Fontan, J.-M., P. Hamel, R. Morin and E. Shragge (2003) Développement économique communautaire dans quatre métropoles, *Organisations et territoires*, Vol.12, No.2, pp.71-77.

Garreau, J. (1991) *Edge City. Life of the New Frontier*. New York, Doubleday.

Germain, A. and G. Rose (2000) *Montréal. The Quest for a Metropolis*. London, John Wiley and Sons.

Hamel, P. (2003) Participation, consultation publique et enjeux urbains. Le cadre du débat public à Montréal et son évolution. In J.-M Fourniau, L. Lepage and L. Simard (eds) *Débat public et apprentissage*, Paris, L'Harmattan (forthcoming).

Hamel, P., H. Lustiger-Thaler and M. Mayer (eds) (2000) *Urban Movements in a Globalising World*. London, Routledge.

Harloe, M. (2001) Social Justice and the City: The New 'Liberal Formulation', *Communication à la conference RC 21*, Amsterdam, June.

Jessop, B. (1997) The Entrepreneurial City. In N. Jewson and S. MacGregor (eds) *Transforming Cities. Contested Governance and New Spatial Divisions*, London, Routledge, pp.28-41.

Jessop, B. (2000) Globalisation, entrepreneurial cities and the social economy. In P. Hamel, H. Lustiger-Thaler and M. Mayer (eds) *Urban Movements in a Globalising World*, London, Routledge, pp.81-100.

Jouve, B. and C. Lefèvre (1999) Introduction. In B. Jouve and C. Lefèvre (eds) *Villes, Métropoles. Les nouveaux territoires du politique*, Paris, Anthropos, pp.9-44.

Kübler, D., B. Schwab, D. Joye and M. Bassand (2002) *La métropole et le politique. Identité, services urbains et citoyenneté dans quatre agglomérations en Suisse*. Rapport de recherche, Lausanne, LaSUR, École Polytechnique Fédérale de Lausanne.

Knox, P.L. (1997) Globalization and Urban Economic Change, *The Annals of the American Academy of Political and Social Science*, No.551, pp.17-43.

Krätke, S. (1992) Villes en mutation. Hiérarchies urbaines et structures spatiales dans le processus de restructuration sociale: le cas de l'Allemagne de l'Ouest, *Espaces et Sociétés*, No.66-67, pp.69-98.

Ku, A. S. (2002) Beyond the Paradoxical Conception of 'Civil Society without Citizenship, *International Sociology*, Vol.17, No.4, pp.529-548.

Laclau, E. and C. Mouffe (1985) *Hegemony and Socialist Strategy. Towards a Radical Democratic Politics*. London, Polity Press.

Lapeyronnie, D. (1999) La ville en miettes, *La Revue du MAUSS*, No.14, pp.19-33.

Lefebvre, H. (1970) *La révolution urbaine*. Paris, Gallimard.

Lefèvre, C. (1998) Metropolitan Government and Governance in Western Countries: A Critical Review, *International Journal of Urban and Regional Research*, Vol.22, No.1, pp.9-25.

Lemire, M. (2000) Mouvement social et mondialisation économique: de l'AMI au cycle du millénaire de l'OMC, *Politique et Sociétés*, Vol.19, No.1, pp.49-78.

Liebs, C.H. (1985) *Main Street to Miracle Mile*. Boston, Little, Brown and Company.

Logan, J. and H. Molotch (1987) *Urban Fortunes: the Political Economy of Place*. Berkeley, University of California Press.

Machimura, T. (1998) Symbolic Use of Globalization in Urban Politics in Tokyo, *International Journal of Urban and Regional Research*, Vol.22, No.2, pp.167-193.

Mayer, M. (2000) Urban social movements in an era of globalisation. In P. Hamel, H. Lustiger-Thaler and M. Mayer (eds) *Urban Movements in a Globalising World*, London, Routledge, pp.141-157.

OECD (2000) The reform of metropolitan governance, *The OECD Policy Briefs*, www.oecd.org/publications/Pol_brief/: 7 p.

Offe, C. (1997) Les liens et les freins : Habermas et les aspects moraux et institutionnels d'une 'autolimitation intelligente. In C. Offe, *Les démocraties modernes à l'épreuve* (textes réunis et présentés par Y. Sintomer et D. Le Saout), Paris, L'Harmattan, pp.168-1998.

Perry B. and A. Harding (2002) The Future of Urban Sociology: Report of Joint Sessions of the British and American Sociological Associations, *International Journal of Urban and Regional Research*, Vol.26, No.4, pp.844-853.

Phillips, A. (2000) Espaces publics, vies privées. In V. Mottier, L. Sgier, T. Carver and T.-H. Ballmer-Cao (eds) *Genre et politique. Débats et Perspectives*, Paris, Gallimard, pp.397-454.

Pickvance, C. (1985) The rise and fall of urban movements and the role of comparative analysis, *Environment and Planning D: Society and Space*, Vol.3, No.1, pp.31-53.

Procacci, G. (2001) Governmentality and Citizenship. In K. Nash and A. Scott (eds) *The Blackwell Companion to Political Sociology*, Oxford, Blackwell, pp.342-351.

Purcell, M. (2000) The Decline of the Political Consensus for Urban Growth: Evidence from Los Angeles, *Journal of Urban Affairs*, Vol.22, No.1, pp.85-100.

Roth, R. (1998) Urban Citizenship – A Contested Terrain, *Communication au XIV° Congrès Mondial de Sociologie*, Association Internationale de Sociologie, Université de Montréal, Montréal, 26 juillet – 1er août.

Roth, R. (2000) New social movements, poor people's movements and the struggle for social citizenship. In P. Hamel, H. Lustiger-Thaler and M. Mayer (eds) *Urban Movements in a Globalising World*, London, Routledge, pp.25-44.

Sancton, A. (2001) Canadian Cities and the New Regionalism, *Journal of Urban Affairs*, Vol.23, No.5, pp.543-555.

Sandercock, L. (1998) *Towards Cosmopolis*. Chichester, John Wiley and Sons.

Sassen, S. (1999) Cracked Casings: Notes toward an Analytics for Studying Transnational Processes,. In J.L. Abu-Lughod (ed) *Sociology for the Twenty-first Century. Continuities and Cutting Edges*, Chicago, University of Chicago Press, pp.134-145.

Sassen, S. (2000) New frontiers facing urban sociology at the Millennium, *British Journal of Sociology*, Vol.51, No.1, pp.143-159.

Savitch, H.V. and P. Kantor (2002) *Cities in the International Marketplace. The Political Economy of Urban Development in North America and Western Europe*, Princeton, Princeton University Press.

Schnapper, D. (1994) *La communauté des citoyens. Sur l'idée moderne de nation*. Paris, Gallimard.

Urry, J. (1998) Globalisation and Citizenship, *Communication au XIV° Congrès Mondial de Sociologie*, Montréal, Association Internationale de Sociologie, Université de Montréal, Montréal, 26 juillet – 1er août, Department of Sociology, Lancaster University, http://www.comp.lancaster.ac.uk/sociology/sco009ju.html

Wolman, H. (1995) Local Government Institutions and Democratic Governance. In D. Judge, G. Stoker and H. Wolman (eds) *Theories of Urban Politics*, London, Sage, pp.134-159.

Chapter 4

Partnerships and Transformation of the State in Urban Britain

Gordon Dabinett

Introduction

The purpose of this Chapter is to examine the role of partnership in national urban policy in England over the last 25 years. During this period there has been a change in political control of the UK national government, with the previous 18 years of Conservative hegemony ending in 1997 with the election of a New Labour Government. A Government that purported to have an alternative model of politics and public policy, based on a Third Way (see Giddens 1998). This Chapter assesses the concept of partnership within this significant break point in the attempts by national political interests to transform the state in Britain. The Chapter draws out critical issues with respect to decentralization, democratic accountability and citizenship in an attempt to explore the scope and construction of policy legitimacy. This is done by assessing the rationale and delivery of urban policy in the periods before and after 1997. It concludes with an examination of the degree to which partnership has become a desired outcome in its own right, an expression of a desired model of politics, rather than a means of securing other political goals, in particular the assertion of individual and market behaviour over the state.

Assessing Partnerships in Urban Policy

Given that the majority of the UK population lives in cities and towns with a population of over 20,000 people (see Table 4.1), it is perhaps not entirely notable that urban policy has been a consistent concern of national governments for over fifty years. Although policy has found articulation in different discourses, it has also retained a number of common concerns (HMG 1977; Department of the Environment 1988; HMG 2000). Past urban policy has sought to achieve broadly similar economic, social, physical and environmental objectives, although the emphasis placed on each of these certainly varied between individual governments (Atkinson and Moon 1994). The expression of spatial targeting also evolved, with the early emphasis on the inner city transfiguring into wider concerns for geographical urban communities such as social housing estates, and for areas of industrial restructuring such as former coalfield or fishing towns. The discourse of disadvantage similarly exhibited subtle changes in language and expression as policy sought, in turn, to deal with poverty in the late 1960s, multiple-deprivation in the 1970s, and social exclusion since the 1990s.   Consistently over this period, national governments articulated broad desires to address uneven spatial development, urban de-investment and physical dereliction, selective population migration, and the geographical concentration of social and economic deprivation. Subsequently, the policy responses exhibited a number of common characteristics:

Table 4.1   Urban Living in a Rural Land – Distribution of Population in England, 1991

Population size of urban area.	Total population (millions)	Cumulative percentage of population (%)	Area covered (Hectares)
Over 250,000	21.8	46.3	509,000
100,000 – 250,000	5.4	57.7	139,000
50,000 – 100,000	4.1	66.5	109,000
20,000 – 50,000	3.8	74.5	105,000
10,000 – 20,000	2.7	80.3	78,000
5,000 – 10,000	2.1	84.8	61,000
3,000 – 5,000	1.2	87.3	39,000
Under 3,000 and rural areas	5.9	100	12,002,000
TOTAL ENGLAND	*47.1*	*100*	*13,042,000*

Source: Urban Task Force (1999) p.29.

- Urban policy was reactive and problem oriented, often delivered through time-limited experiments or initiatives, rather than being a mainstream service or sectoral policy.
- Urban policy targeted specific additional resources on small geographical areas, involving a disparate set of projects, within multi-annual programmes, seeking to achieve multiple objectives.
- Urban policy found expression in a national response, but was always implemented within a multi-level government structure, that cast central-local relations within a predominantly centralized state.

Whilst it is possible to identify the common ground of the urban problem as a rationale for government intervention, it must be fully acknowledged that urban policy reflects and is a consequence of deep-seated inequalities in society. Urban policy discourses often focus on how cities and urban areas might be successfully developed, but this perspective must not hide more overtly distributional agendas. Notions of uneven spatial development, the distribution of resources, and the origins of policy in responses to racism and urban unrest, all point towards elements of social control and the exercising of state power within the implementation of this national policy. Given the potential dominance of national government in the formulation of policy, this Chapter takes a statist view of UK urban policy. However, complex political and governance issues are also revealed in the relationship between what urban policy sought to achieve, and how it was delivered or implemented. One explanation of the urban problem and the crisis within our cities throughout the chronology UK urban policy has been the notion of institutional failure. This is a common theme within neo-liberalist, social-democratic and more radical views on urban change, but inevitably is given different meanings and generates different policy responses in practice. Consistent within these arguments and practice are political rather than managerial notions of democratic accountability and, increasingly, the empowerment of disadvantaged communities and citizens.

Institutional behaviour in cities has seen a supposed change from urban government to urban governance over the twenty-five year period of UK urban policy (Stoker 1999). Until the 1980s cities were managed by democratically elected local authorities. These multi-purpose authorities were given their powers, responsibilities, functions, resources and various degrees of discretion and autonomy by national government. Whilst the relationship between national government and urban local authorities changed in detail over time, significant shifts began to occur during the 1980s, as cities that were generally controlled by Labour party politicians came into conflict with a neo-liberal Conservative national government. Over this decade a number of fundamental changes in the way cities were governed occurred as a result of national government policies. Urban local authorities had functions removed or reduced, had to deliver services within new implementation systems and structures, and had to work with new agencies and quangos set up to deliver national government services and functions. In addition, the capacity of local authorities to steer was reduced through national government guidance, and they were compelled to enter in to new institutional interactions, most notably partnerships. It is against this background that we need to examine and assess the role and meaning to be attached to partnership and governance in urban regeneration.

Reviews of urban regeneration practice have shown that partnership working within the UK has always been a concept associated with inner cities and urban regeneration (Bailey 1995). However this observation can equally obscure specific institutional practices and varied philosophies, since the meaning of partnership can contain a high level of ambiguity with the delimitation of public and private sectors requiring careful consideration (Macintosh 1992). Studies of partnerships in wider UK welfare policies found that most commentators emphasize the difficulty of definition; in particular the distinction between partnerships and other types of inter-organizational relationships was identified as being problematic (Clarke and Glendinning 2002). A further but significant input to the discussions in this chapter is also the view that partnerships are often the benign products of empty political rhetoric (Fairclough 2000).

The common rationale for partnerships advocated within much of the policy and administrative literature focuses on co-ordination, to deal with the longstanding issue of achieving policy efficacy between and within levels of government. In contrast, writers about urban partnerships have generally indicated concerns with a wider range of processes, such as budget enlargement, the creation of synergies and adding value through skills acquisition and knowledge transfer. A working definition can be found, where urban regeneration partnerships are seen to be the mobilization of a coalition of interests drawn from more than one sector in order to prepare and oversee an agreed strategy for the regeneration of a defined area (Bailey 1995). This perspective is used in this chapter: partnership is seen to involve relationships between two or more different sectors within a mixed economy, including public-public, public-voluntary, public-community and public-private.

Of particular relevance to the arguments developed in this assessment of UK urban policy is the notion of transformative partnerships. The transformation model emphasizes changes in the aims and cultures of the partners' organizations. Hastings (1996) developed the framework originally put forward by Mackintosh (1992) to expand issues of power and attitudes to develop models of uni-directional transformation or mutual transformation. In the former, a battle for change is joined which involves an unequal power relation, in which, crucially, one or all partners are unwilling to change. The later is characterized by reciprocal challenges to pre-existing culture, and less coercive, antagonistic relationships. Each partner might be willing to accept to change, as well as aspire to change others, and there may be a desire to learn as well as to teach. Such a perspective raises key questions of who is involved in a partnership, and what roles and powers they have, but also asks who gives authority and legitimacy to a partnership.

As indicated earlier, partnerships in UK urban policy have a relatively long history, but the notion of these arrangements serving the transformation of the state, as opposed to being a simple means of implementing policy, was to have particular resonance in the 1980s. Before this decade, strategic partnerships in the main urban areas were based on a relationship between central and local government: 'by which urban change could be managed through improved co-ordination between the two tiers of government and the channelling of additional resources to areas of greatest need' (Bailey 1995, p.225). Although the mechanisms for funding such partnerships were initially retained by the incoming Conservative Government of 1979, they soon became sidelined in favour of the new political project of promoting the enterprise culture (Deakin and Edwards 1993) and a public-private partnership model became dominant (Barnekov, Boyle and Rich 1989). Consequently, it is necessary to outline the development of urban partnerships in this critical period of the 1980s, as well as exploring the period associated with New Labour and the Third Way after 1997.

Entrepreneurial, Competitive and Consensual Urban Policy

The Conservative victory of May 1979 brought radical change to the UK. Margaret Thatcher's government was committed to 'rolling back the frontiers of the State and improving the functioning of the market economy' (Lawson 1992, p.52). The subsequent consecutive four terms of office (1979-1997), with substantial majorities in the Houses of Parliament on each occasion, enabled the national government to pursue their policies over an eighteen-year period that resulted in a new political hegemony. This hegemony rejected the previous Keynesian welfare state, and instead promoted a market-driven economy, a reduction in the size of the public sector and the extension of privatization, the lowering of taxes, and the maintenance of low inflation as a central macro-economic goal. These events had their origins in the severe economic crises faced by the country during the 1970s (see Tables 4.2 to 4.4), and whatever the diagnosis of the British economy's ills, this period clearly represented a paradigmatic shift that has been termed 'restructuring' (Green 1989). Restructuring not only involved a change in the composition of industries and in the labour force but, more generally, an alteration in the processes and structures within which capital accumulation occurred. Furthermore, the new Government of 1979 was convinced that reform and restructuring of the economy could not be carried out without an approach based on conflict. In practice this resulted in 'populist' policies that divided the working class (Desai 1989), created an underclass dependent on transfer payments from the state, and dismantled the traditional corporatist forms of state-trade union-business relations at the national level (Bassett 1996). Economic restructuring and the pursuit of a free-market ideology were both to have critical and far-reaching impacts on the urban areas of the UK.

Table 4.2   The Urban Exodus – Population Change in British Cities 1951-1997

(Population millions)	1951	1981	1991	1997	*(%change 51-81)*	*(%change 81-91)*	*(%change 91-97)*
Conurbations[1]	19.4	17.5	17.2	17.4	*(-9.8)*	*(-1.7)*	*(+1.4)*
Rest of Great Britain	29.4	36.1	37.7	38.6	*(+22.5)*	*(+4.4)*	*(+2.5)*

Source: Adapted from Begg, Moore and Altunbas (2002).

Urban areas, and in particular the large conurbations, were to suffer disproportionately from the economic consequences of this restructuring (see Tables 4.3 and 4.4). Not only did they loose a large number of manufacturing jobs, but the attack on public sector services also had the double impact of cutting welfare support and reducing employment opportunities for many inner city residents. During the 1980s cities became the sites of social conflict and rioting, and some large Labour controlled urban local authorities, including London, Liverpool, Manchester and Sheffield, attempted to pursue alternative local policies in direct opposition to the Conservative national government (Mackintosh and Wainwright 1987). Any period of restructuring, as represented by events of the 1980s, is a situation where different groups will vie for power, and the process is clearly open to political determination. However, despite restructuring becoming a global phenomenon to which different nation states adopted a variety of responses at this point of time, 'Thatcherism' attempted to promote its policies as inevitable and the only way out of the crisis – there is no alternative. This philosophy of the 'free market and the strong state' (Gamble 1994) was illustrated when, in the mid-1980s, the government imposed

[1] The seven conurbations of Glasgow/Clydeside; Newcastle/Tyne and Wear; Liverpool/Merseyside; Greater Manchester; Leeds/West Yorkshire; Birmingham/West Midlands; London.

further financial restrictions on all local authorities and abolished the six Labour controlled metropolitan authorities (the Tyne and Wear, Merseyside, West Midlands, West Yorkshire, South Yorkshire and Greater London Councils).

Table 4.3   The Urban Economic Crisis – Employment Change in British Cities 1959-1997

(Employment millions)	1959	1981	1991	1997	(%change 59-81)	(%change 81-91)	(%change 91-97)
Conurbations[1]	10.2	8.5	8.4	8.8	(-16.4)	(-0.6)	(+3.8)
Rest of Great Britain	12.5	14.7	16.2	17.0	(+18.1)	(+10.1)	(+5.0)

Source: Adapted from Begg, Moore, and Altunbas (2002).

Table 4.4   Social Costs of Restructuring – Unemployment Change in British Cities 1951-1997

(Unemployment thousands)	1951	1981	1991	1997	(%change 51-81)	(%change 81-91)	(%change 91-97)
Conurbations[1]	199.9	950.5	973.8	781.4	(+375.5)	(+2.4)	(-19.8)
Rest of Great Britain	275.4	1,972.7	1,508.3	1,138.7	(+616.2)	(-23.5)	(-24.5)

Source: Adapted from Begg, Moore, and Altunbas (2002).

However, far wider and longer lasting effects on urban governance were to be prompted by the pursuit of new public management (Pollitt 1990). In essence, where services could not be provided directly through the market by privatization, such as the railways and electricity supply, then 'quasi-market' regimes were enforced. For example, local authorities were required to put some services out to tender to the private sector (Compulsory Competitive Tendering) whilst other in-house services were subjected to internal markets. A substantial amount of the social housing stock in urban areas was sold-off to tenants under the 'Right-to-Buy' legislation. Furthermore, education budgets were removed from the control of local authorities and given to individual school governing bodies, and state funding for the training of post-16 year olds was allocated and managed by business-led partnership agencies – Training and Enterprise Councils.

What was particularly distinctive about the 1980s was the extent to which under Thatcher the 'use of the market was about power as well as efficiency' (Hill 2000, p.124). Emphasis was placed on reducing local government autonomy and

---

[1] See Previous Footnote 1.

introducing a profusion of non-elected, private-sector-led bodies that largely bypassed local authorities. A centralization of powers was achieved through increasingly prescriptive constraints on local government functions and finances. The influence of social democratic structures was reduced, whilst increasing that of parents, house owners, consumers and private businesses. Thus central government policy 'sought structurally to privilege individuals who, traditionally in the UK, have not been involved in the mechanisms of local governance' (Ward 1997, p.1496). National urban policy frameworks provided a particular stimulus to this development, as partnerships were used in cities as a vehicle for furthering local entrepreneurialism and the attack on the local state (Deakin and Edwards 1993).

On coming to office in 1979 the Conservative government appeared to have little interest in urban policies, but riots and racial tensions in the early 1980s drew stark attention to conditions in the inner cities and placed urban issues on the political agenda. In general terms, the post-1979 Conservatives believed that the problems of British cities derived from the flight of private capital, largely caused by rigid local government bureaucracies and inflexible labour markets (Atkinson and Moon 1994; Lawless 1989). The need was to encourage investors to return. Urban Development Corporations (UDCs) and Enterprise Zones (EZs) were the most innovatory attempt to facilitate free-market, property-led urban regeneration during the years of Margaret Thatcher's leadership (1979-1990). Between 1981 and 1985 23 EZs were designated, mainly located in run-down urban areas (see Table 4.5). Since they reduced certain regulations on businesses as well as providing tax and other financial incentives, EZs were equally regarded as laboratories for testing free-market principles. However, by 1988 the UDCs had become regarded as the 'most important attack ever made on urban decay' (Michael Heseltine, Secretary of State for the Environment, quoted in Imrie and Thomas 1999). After the initial declaration of the London and Merseyside UDCs in 1981, their number had increased to thirteen by the end of 1992 with substantial expansion in numbers and budget taking place after 1987 (see Table 4.6).

UDCs were given the task of regenerating designated areas within the major cities of England and Wales (see Table 4.6). They were supposedly set up in the national interest, thereby justifying the taking of large areas of land out of the control of local authorities, many of which were controlled by the Labour Party, and placing them in the hands of a non-elected local body, the UDC. Financially and administratively they were accountable to Central Government. The objective of each UDC was to secure the physical and economic regeneration of their designated area. UDCs had a variety of powers to assemble and prepare land but most crucially they became the local planning body, usurping the elected local authority in this role. Each UDC was managed by a Board appointed by Central Government. Most members and the chair were from private businesses. Local authorities had a minority presence, and usually there was no role for community or residents interests.

Table 4.5   Enterprise Zones[2] – Estimated Total Public Costs and Land Developed 1981-1986

£millions, 1985/86 constant prices	1981/82	1982/83	1983/84	1984/85	1985/86	Total 1981-86
Rate Relief	*6*	*13*	*15*	*21*	*27*	*82*
Capital Allowances	*20*	*20*	*20*	*30*	*60*	*150*
Land Acquisition by Public Agencies	*0*	*8*	*11*	*9*	*1*	*29*
Infrastructure and Reclamation Expenditures	*11*	*24*	*36*	*34*	*30*	*135*
TOTAL COSTS	37	65	82	94	118	396
Land Developed (Hectares)	688	907	1,369	1,843	2,000	6,807

Source: Department of the Environment (1987) p.11.

Over time, these arrangements reinforced a more general shift towards property led urban renewal (Healey et al. 1992), based on inter-locality competition and local 'boosterism' (see Ashworth and Voogd 1990). According to Lawless (1996), although this notion emerged from a number of ideological and political roots, the influence of the United States was considerable. There was a long tradition of overtly pro-business urban governance in the US where 'local partnerships have assumed a kind of neutral local corporatism founded on the assumption that a pro-business programme has to be good for a city and its inhabitants' (Lawless 1996, p.36). However unlike North America, the regimes in the UK were predominantly a nationally prescribed response (Ward 1997), inevitably linked to the symbolism of high-profile property development, the trickle-down of benefits, and the privileging of private-sector interests over those of local communities.

[2] Twenty-three enterprise zones: 10 designated in 1981/82 and 13 designated in 1983/84.

Table 4.6   Urban Development Corporations 1981-1998

Urban Development Corporation	Year of Designation	Size of Development Area (Ha)	Reclaimed Land 1997 (Ha)	Grant Aid (£mill)	Private Sector Investment (£mill)
London Docklands	1981	2,150	776	1,860.3	6,505
Merseyside	1981	960	382	385.3	548
Black Country	1987	2,598	363	357.7	987
Cardiff Bay	1987	1,093	310	370.0	774
Teesside	1987	4,858	492	350.5	1,004
Trafford Park	1987	1,267	176	223.7	1,513
Tyne and Wear	1987	2,375	507	339.3	1,115
Bristol	1988/9	420	69	78.9	235
Sheffield	1988/9	900	247	101.0	686
Central Manchester	1988/9	187	35	82.2	373
Leeds	1988/9	540	68	55.7	357
Birmingham Heartlands	1992/3	1,000	115	39.7	217
Plymouth	1992/3	70	10	44.5	41
Total				*3,918.8*	*14,320*

Source: Adapted from Imrie and Thomas (1999) pp.13-27.

A new phase of policy was ushered in by the launch of the innovative urban policy mechanisms, City Challenge in 1991 and 1992 and the Single Regeneration Budget (SRB) from 1994. These saw a slow shift in favour of a more people-based approach, which began to recognize the importance of social and housing investments. These events occurred at a point of time when urban property markets had collapsed and there was a change of leadership within the Conservative Party (and therefore Premiership) to John Major (1990-1997). National policy still placed emphasis on promoting local entrepreneurialism, but the emphasis on private-sector leadership was tempered as local government was reintroduced within a new form of competitive urban policy frameworks (Haughton and While 1999). This was heralded by Government as a new approach that could give local people more influence, but it had two critical features:

- The Partnership had to submit a competitive bid to central government, with winners and losers in each annual round (see Table 4.7 for City Challenge).
- Successful Partnerships had to implement a multi-annual programme in their selected areas by grant funding projects, but any local flexibility was constrained by central government guidance and targets, a contract culture (Oatley 1998).

Table 4.7 Outcomes of City Challenge 1991 and 1992

CITY CHALLENGE 1991
Winning Bids by Local Authority:

1.	Lewisham, London	2.	Tower Hamlets, London
3.	Bradford, West Yorkshire	4.	Nottingham, East Midlands
5.	Dearne Valley, South Yorkshire	6.	Newcastle, Tyne and Wear
7.	Liverpool, Merseyside	8.	Wirral, Merseyside
9.	Manchester, Greater Manchester	10.	Wolverhampton, West Midlands
11.	Middlesborough, Teesside		

Rejected Bids by Local Authority:

1.	Birmingham, West Midlands	2.	Salford, Greater Manchester
3.	Bristol, Avon	4.	Sheffield, South Yorkshire

CITY CHALLENGE 1992
Winning Bids by Local Authority:

1.	Barnsley, South Yorkshire	2.	Lambeth, London
3.	Birmingham, West Midlands	4.	Leicester, East Midlands
5.	Blackburn, Lancashire	6.	Newham, London
7.	Bolton, Greater Manchester	8.	North Tyneside, Tyne and Wear
9.	Brent, London	10.	Sandwell, West Midlands
11.	Derby, East Midlands	12.	Sefton, Merseyside
13.	Hackney, London	14.	Stockton, Teesside
15.	Hartlepool, Teesside	16.	Sunderland, Tyne and Wear
17.	Kensington and Chelsea, London	18.	Walsall, West Midlands
19.	Kirklees, West Yorkshire	20.	Wigan, Greater Manchester

Rejected Bids by Local Authority:

1.	Bradford, West Yorkshire	2.	Middlesborough, Teesside
3.	Bristol, Avon	4.	Newcastle, Tyne and Wear
5.	Burnley, Lancashire	6.	Nottingham, East Midlands
7.	Coventry, West Midlands	8.	Oldham, Greater Manchester
9.	Doncaster, South Yorkshire	10.	Plymouth,
11.	Dudley, West Midlands	12.	Preston, Lancashire
13.	Gateshead, Tyne and Wear	14.	Rochdale, Greater Manchester
15.	Greenwich, London	16.	Rotherham, South Yorkshire
17.	Halton, Merseyside	18.	St Helens, Lancashire
19.	Hammersmith and Fulham, London	20.	Salford, Greater Manchester
21.	Haringey, London	22.	Sheffield, South Yorkshire
23.	Hull, Humberside	24.	South Tyneside, Tyne and Wear
25.	Islington, London	26.	Southwark, London
27.	Knowsley, Merseyside	28.	Tower Hamlets, London
29.	Langbaurgh, Teesside	30.	Wandsworth, London
31.	Leeds, West Yorkshire	32.	Wolverhampton, West Midlands
33.	Liverpool, Merseyside	34.	The Wrekin, Merseyside

Source: Based on Oatley and Lambert in Oatley (1998) p.112.

As a result, it was argued that these new regimes involved a new style of corporatist alliance between central and local government, the private sector and voluntary and community organizations. This alliance was subjected to constraints imposed by The Treasury and other central government departmental procedures (Oatley 2000). Evidence suggested that many partnerships were unlikely to happen naturally and central government sought to enforce this style of working, and as a result failed 'to devolve power and instead reinforced the fragmentary policies that underpin the British government's neoliberal agenda' (Ward 1999, p.1493). The concept of partnership in urban policy therefore grew in importance over the two decades of Conservative rule. While the traditional boundaries between the public and private sectors fluctuated back and forth, Bailey (1995) suggests that a policy space was created for a series of experiments in urban governance, a space characterized by contestation.

In contrast, the post-1997 UK Government has emphasized a collaborative discourse. Partnership is a key word for New Labour (Fairclough 2000), and sees it as an essential part of its more general way of governing (Clarke and Glendinning 2002). The urban agenda was also in the vanguard of the policy and legislative programmes of the new government, with the main initiatives being launched in the first term of office between 1997 and 2001. The most explicit approach to urban areas was developed as a result of the Government setting up an Urban Task Force in 1998 (Urban Task Force 1999), and many of its suggestions for an urban renaissance were subsequently incorporated in to the Government's White Paper – Our Towns and Cities (HMG 2000). The White Paper contained numerous proposals for action, and some innovative interventions such as the use of new fiscal measures. One of the clearest expressions of new policy was the establishment of urban regeneration companies (URC). By 2002 eleven URCs had been established to operate within designated areas within a number of urban local authorities (See Table 4.8). The Government claimed that URCs represented a major new development in regeneration policy. 'They are a recognition of the failure of the market mechanism in many areas of our towns and cities and of the inability of past public sector interventions to correct these failures and create lasting improvements' (Department of Transport, Local Government and the Regions 2001). Although it might be too early to judge the outcomes of these new policies, doubts might be legitimately raised about the extent to which there has been a significant policy change. However, they represent some change from UDCs. Although they are also managed by Central Government appointed boards that are accountable to Central Government departments, unlike UDCs these are stakeholder partnerships. They are made up of the local authority, private sector interests, the new regional development agencies (set up by Labour), and English Partnerships (the national government property development agency). In addition, unlike UDCs, they have no special powers or funding, but rely on the combined functions and resources of these stakeholders.

Table 4.8   Spatial Targeting in New Labour's Urban Policy, 2002

39 New Deal for Communities Areas

East Manchester	Aston, Birmingham	Devonport, Plymouth
Shoreditch, London	South Kilburn, London	Old Heywood, Rochdale
Kings Norton, Birmingham	Wood End, Coventry	Charlestown, Salford
Barton Hill, Bristol	Derwent, Derby	Burngreave, Sheffield
Braunstone, Leicester	Doncaster Central	Thornhill, Southampton
Aylesbury Estate, London	North Fulham, London	Hendon, Sunderland
West Ham, London	Seven Sisters, London	Bloxwich East, Walsall
East Brighton	West Central Hartlepool	All Saints, Wolverhampton
Preston Road, Hull	Finsbury, London	Huyton, Knowsley
Kensington, Liverpool	Clapham Park, London	Greets Green, Sandwell
West Gate, Newcastle	New Cross Gate, London	West Middlesborough
North Earlham, Norwich	Marsh Farm, Luton	Radford, Nottingham
Ocean Estate, London	Hathershaw, Oldham	Little Horton, Bradford

11 Local Authorities with Urban Regeneration Companies

Sheffield	Bradford	Corby
Manchester	Tees Valley	Leicester
Liverpool	Sunderland	Camborne, Pool and
Hull	Swindon	Redruth, Cornwall

Source: Various.

Without doubt though, one of the most significant changes in urban policy came with the emphasis that New Labour placed on reducing social exclusion as a central element of all government's activities. Initially this saw the establishment of the Social Exclusion Unit (SEU), which was a cross-departmental team set up within the machinery of national government administration and policy making (HMG 1998). Subsequently, a Neighbourhood Renewal Unit (NRU) was set up in 2001 within central government to implement a national strategy and to oversee the spending of the Neighbourhood Renewal Fund (Neighbourhood Renewal Unit 2002).   The Renewal Fund was allocated to the eighty-eight local authorities in England containing the highest levels and concentrations of socially excluded populations, many of which were in the main urban areas.   A more explicit urban response was the spatial targeting of thirty-nine neighbourhoods to benefit from a purportedly radical programme, the New Deal for Communities (NDC) (see Table 4.8). Each area of approximately 4,000 households was allocated some £50 million over ten years to address poor health, low educational and skills levels, high joblessness and rates of crime, and under-investment in housing and the physical environment (Neighbourhood Renewal Unit 2001). The NDC programme was announced in 1998 and was intended to bridge the gap between the most deprived neighbourhoods and the rest of the country. The NDC initiative falls within a long-standing strand of policy innovations often referred to as area-based initiatives (ABIs). In some respects NDC paralleled earlier ABIs, such as City Challenge, but New Labour claimed that the NDC programme placed greater emphasis on involving and engaging local residents.

Therefore, despite the many claims to associate New Labour with the Third Way, Allmendinger and Tewdwr-Jones (2000) believe post-1997 urban policy cannot be 'neatly boxed in a hermetically sealed ideological family' (p.1385), and the overall approach appeared more practical than theoretical, often leading to potentially contradictory elements. One illustration of this contradiction was articulated in urban policy, as cities were seen as the source of economic opportunity and a key asset in the securing of national economic goals, but were also seen as localities of the worst concentrations of social exclusion. New Labour argued that this contradiction could be resolved through the mainstreaming of national policy. Purportedly, the successes and best practices generated by ABIs and other specific measures would be adopted and put into practice more widely by bending the funding and activities of the main service providers and public agencies. This argument reaffirms the need to consider wider issues of urban governance contained in the new urban policy and, in particular, partnership working.

State Transformation and Urban Policy

The nature of state transformation articulated in the discourses and practices of UK urban policy can be assessed by critically examining public-private relations, central-local government relations, and the empowerment of communities. Previous Conservative urban policy was clearly characterized by privatism, where the compulsory requirement of partnership working was perceived by central government as a means to enforce a role for private sector interests, most notably in urban development corporations (Imrie and Thomas 1999). From one perspective, the increased influence given to business interests can be seen as an attempt to transform urban governance by exposing public agencies to competition and business management practices. Certainly the emergence of private sector influenced growth coalitions within the political behaviour of many cities would indicate such transformative outcomes (Bassett 1996; North, Valler and Wood 2001). However, other evidence is more circumspect about the actual impact of strategic public-private partnerships (Peck 1995; Raco 1997), arguing business representatives often remained peripheral to public sector dominated decision-making. Davies (2002) argues that an 'ideological commitment to partnership between local political and business elites existed to a greater or a lesser extent in all areas...but business participation was symbolic' (p.178). Very often, such arrangements represented grant coalitions only, existing to ensure urban areas were successful in their bids for national government and European Union regeneration funds. However, a widening of the interpretation placed on such partnerships reveals that urban programmes were also associated with a market-oriented ideology that linked an increasing emphasis upon the promotion of private sector investment with the targeting of tightly restrained public resources and the introduction of competition within urban services. This general transforming effect is reflected upon by Oatley (1998), claiming that urban policy before 1997 undermined local democracy, but also contributed to the restructuring of the relation between state and civil society, as exhibited in the continually contested territory of relationships between the national and local tiers of government.

The early rhetoric of New Labour certainly rejected this seemingly ideologically driven practice of the previous Governments, with Tony Blair claiming to 'reject the rampant laissez-faire of those who believe government has no role in a productive economy.' (quoted in Driver and Martell 1998).   In terms of public-private relationships, the post-1997 urban policies were to exhibit severe contradictions or high levels of pragmatism.   On one hand, the regional development agencies, set up by the Labour government in 1998 to oversee the spending of SRB and other regeneration funds, were led by Government appointed boards with private sector representation and chairmen from regional businesses, and were very similar in structure to the earlier UDCs.   On the other hand, the URCs were partnerships between local, regional and national public agencies or quangos, but continued the use of non-elected bodies to implement urban policy through supply-side interventions in urban land and property markets.   Whilst these responses might give the appearance that New Labour was stepping back from the transformative agenda pursed by the previous conservative governments, elements of privatism remained in place.   New Labour did not set about on a re-nationalization programme, and instead significantly increased the use of the private funding initiative (PFI) and public-private partnerships (PPPs) to provide urban public infrastructure such as schools, hospitals and transport infrastructure (Driver and Martell 2002; Fairclough 2000).

Privatism under the Conservatives during the 1980s had seen not only the promotion of private sector interests, but also direct attempts by national government to limit and control the powers and influence of urban local authorities.   The introduction of new public management principles (see Pollitt 1990) had led to a great deal of institutional churning and fluidity as quangos and various multi-agency partnerships were established to implement urban regeneration policies. However this proliferation of institutional arrangements contributed to further governance failure. Consequently the lack of integration within the machinery of government gave reason for the Labour Government to set up inter-departmental groups (see Regional Co-ordination Unit, Social Exclusion Unit, and Neighbourhood Renewal Unit), to strengthen regional structures (Government Offices in the Regions, Regional Development Agencies, Regional Assemblies), and to establish Local Strategic Partnerships (LSPs). LSPs were seen as a key element in these reforms, and were set up in each local authority urban area.   They were given a strategic role to bring together public, private and voluntary sector service providers with local communities and businesses.   Their aim is to help co-ordinate national and local initiatives and to cut out duplication and to reduce bureaucracy by simplifying existing partnerships (Neighbourhood Renewal Unit 2002).   Although intended to be inclusive and broad based bodies, their membership and authority still had to be approved by central government.   Indeed, all these arrangements for the state were largely top-down, rather than bottom-up, institutions characterized by networking and trust.   They rested on democratic accountability at a national level only, and there was evidence of an emerging urban elite – managers and power brokers – who exercised power and influence at the heart of the interlocking network of partnership leaders (Stewart

1998). So to what extent was this practice different to the new public management so aggressively pursued by the previous Conservative neo-liberalist hegemony?

Despite the pragmatic and managerialist solutions apparent in these discordant responses, New Labour's approaches to urban governance were also consistently underpinned by two transformative ambitions – devolution and modernization. Certainly the setting up of new forms of territorial government in Scotland, Wales, Northern Ireland and Greater London was an approach that was never supported by the Conservatives, and represented a significant and radical break in the past traditions of governing the UK (Tomaney 2002). The policy towards urban local government is less clear. Whilst the devolution debate was largely steered by political considerations (e.g. national separatists movements), the modernization of local authorities has been dominated by familiar debates from the Conservative period, pertaining to the performance and efficiency of public services. The first New Labour government during 1997-2001 seemed equally determined to limit overall levels of public expenditure by the detailed control and influence of those agencies charged with delivering public services on the ground. Thus many features of new public management appear in new or different guises, such as the setting of targets in public service agreements, the close monitoring of local authority services through Best-Value, and experiments in local accountability, such as local authority Cabinets and elected urban local authority Mayors. Whilst urban policy does not appear to be a critical element in the delivery of this modernization agenda, the success of urban policy was seen in the White Paper to be contingent on these wider state transformations (HMG 2000).

Having attempted to circumvent urban local authorities in the early eighties, the Conservative government then faced the need to create a means of reincorporating local interests in to the development and implementation of urban policy. The introduction of City Challenge in 1991 and the SRB in 1994 represented significant and defining changes in pre-1997 urban policy in this respect. It has already been argued that area-based regeneration programmes that aim to empower communities have featured prominently in urban policy. Empowerment might be seen not as an outcome of a single event, but rather a continuous process that enables people to understand, upgrade and use their capacity to better control and gain power over their own lives. It provides people with choices and the ability to choose, as well as to gain more control over resources they need to improve their conditions (Schuftan 1996). However, the notions of capacity-building or empowerment have been used with varying and contested meanings within urban programmes, and supposedly similar programmes have been promoted as part of considerably differing policy agendas.

Partnerships that incorporated community representatives and actors were particularly valued from the market-oriented perspective. The potential for capacity-building associated with such partnerships was seen to offer a number of desired outcomes. It could contribute to the enhancement of cost-effectiveness by levering in additional resources via self-help mechanisms and fill the potential gaps that might otherwise emerge in the provision of services, as the local state was rolled back. In addition, the active involvement of the voluntary and community sectors offered opportunities to enhance the legitimacy of these arrangements in the eyes of local actors. However, Mayo (1997) points out that such short-term swoops into areas identified as ripe for renewal were unlikely to produce partnerships that were effective, let alone empowering for the differing interests within the communities in question. In reinforcement of this view, Bailey (1995) claims that partnerships placed differing interpretations on whether regeneration was to take place primarily in the interests of local residents, or whether they were merely bringing local knowledge and commitment and ownership to projects that had emerged from technocratic rather than democratic processes. In such technocratic processes, dominant interests such as local authorities and government quangos had already heavily structured the boundaries of debate and action.

Once again, the early rhetoric of New Labour promised much, with an emphasis on people and the involvement of communities and minority groups in many of the implementation measures and agencies. In the Foreword to the Urban White Paper, the Deputy Prime Minister claimed: 'Our guiding principle is that people must come first. Our policies, programmes and structures of governance are based on engaging local people in partnerships for change with strong local leadership' (HMG 2000, p.5). The practice of community involvement in urban partnerships was promoted in a number of ways, but as yet the outcomes of New Labour's emphasis on community involvement, such as NDC, is less than clear. It is still a highly contested area, often characterized by mistrust, misunderstanding and deep seated problems, as illustrated by the re-occurrence of severe social and racial unrest in the spring and early summer of 2001 (Home Office 2002). Clearly such partnership working was being used to potentially change or reinforce the existing power and political relations in urban areas, rather than simply to improve urban governance. There is a great deal of debate and contradictory evidence about the extent to which control is relinquished or simply reshaped, especially where there are budget constraints, an increased bureaucracy and continued central government regulation. But empowering communities cannot be conflict-free, and has to address tensions. Such tensions exist between the need for administrative and financial accountability and the desire for individual flexibility, between inclusive participation and strong leadership, and between a community consensus and socially diverse neighbourhoods. Certainly the resolution of these requires institutional capacity, but is this sufficient without a shift of power?

Referring to the broad-based regeneration partnerships that were emerging in the last years of Conservative national government, Mayo and Anastacio (1999, p.19) observed that:

> 'these policies may also be exacerbating processes of disempowerment, further marginalising the already marginalized – increasing polarisation rather than reducing the democratic deficit. And both of these divergent processes can occur within the context of market-oriented social welfare policies, which area regeneration partnerships may be effectively legitimising.'

Such concerns have been extended to New Labour's community based renewal programmes in the wider context of its social policies, referred to as welfare to workfare (Peck 1999). This criticism has also been linked to the early interest and empathy shown by New Labour for the ideas and ideals held by communitarians, which supposedly redefine social policies and social responsibilities (Etzioni 1993). The communitarian agenda has also been regarded as an idealist moral order of active and responsible citizenship, but one that ignores economic inequalities, class, gender, and race. Certainly a moral agenda has been associated with New Labour's welfare policies that underpin many of the urban interventions, a moral position that is intolerant of non-conforming behaviour, and attaches importance to individual independence as a primary requirement of social inclusion.

Partnerships and New Urban Governance

A broad and a-theoretical definition of partnership working leads to a view that partnerships have been a consistent feature of urban regeneration practice in the UK (Bailey 1995). However, a critical interpretation is more useful, in that this can reveal transforming effects partnerships might have on power relationships and the involvement of political interests in the urban polity. Partnerships in urban policy can reflect wider changes in the values and legitimacy attached to the state. From this perspective, the last twenty years of national urban policy in the UK reveals that partnerships, once used instrumentally by the Conservative Government to effect privatism and to challenge the local state, have now become assimilated within the policies of New Labour to implement a desired political model. Thus the whole discourse of partnership has been used to express 'the non-ideological, non-dogmatic orientation of the Third Way, moving beyond the 'old' ideological commitments to the market or the state. Partnership exemplifies the pursuit of pragmatic solutions to policy problems' (Clarke and Glendinning, p.33). The post-1997 New Labour governments attempted to develop a consensus about the desirable form of urban policy based on a new pragmatism, consisting of a combination of new institutional fixes and a resurrected managerialism (Oatley 2000).

But these compulsory partnerships were clearly also an attempt to recruit dependent local partners into the project of modernizing government. From this perspective, New Labour's partnerships represented another stage in dealing with the contradictory demands and pressures of governing, but in a contingent and unstable way. They represent potentially contested sites of power rather than consensus and the co-ordination of policy delivery. It would appear that it is too early to dismiss the national state as a center of power and authority, and the practice of introducing New Labour's partnerships certainly represents an attempt to reassert control, despite the rhetoric that claims decentralization and empowerment. Perhaps what is happening is not so much a decline of the state as a gradual transformation of the role of the state in urban society, within some new and emerging global regime. The autonomy of national and local state actions is clearly bounded, and contradictions might be seen to emerge from a desire of New Labour to replace a representative democracy with a more deliberative democracy, a form of democracy in which groups, networks and other intermediate social structures, such as partnerships, are used as mechanisms for a form of governance that focuses on the individualism of citizens in the urban polity. The test of New Labour's urban partnerships might be the extent to which the urban civic realm is restored in ten years time, and the creation of a realm that also brings about an increase in the legitimacy and accountability of Government actions for those people and communities at greatest disadvantage, so often portrayed as alienated and disenfranchised from urban politics and political processes.

References

Allmendinger, P. and Tewdwr-Jones, M. (2000) New Labour, New Planning? The trajectory of planning in Blair's Britain, *Urban Studies*, Vol.37, No.8, pp.1379-1402.

Ashworth, G. and Voogd, H. (1990) *Selling the City: Approaches in Public Sector Urban Planning*. London, Belhaven Press.

Atkinson, R. and Moon, G. (1994) *Urban Policy in Britain*. Basingstoke, Macmillan.

Bailey, N. (1995) *Partnership Agencies in British Urban Policy*. London, UCL Press.

Barnekov, T., Boyle, R. and Rich, D. (1989) *Privatism and Urban Politics in Britain and the United States*. Oxford, Oxford University Press.

Bassett, K. (1996) Partnerships, Business Elites and Urban Politics: New Forms of Governance in an English City? *Urban Studies*, Vol.33, No.3, pp.539-555.

Begg, I., Moore, B. and Altunbus, Y. (2002) Long-run Trends in the Competitiveness of British Cities. In I. Begg (ed) *Urban Competitiveness: Policies for Dynamic Cities*. Bristol, The Policy Press, pp.101-134.

Clarke, J. and Glendinning, C. (2002) Partnership and the Remaking of Welfare Governance, in C. Glendinning, M. Powell and K. Rummery (eds) *Partnerships, New Labour and the Governance of Welfare*, Cambridge, Polity, pp.33-50.

Davies, J. (2002) Regeneration Partnerships Under New Labour: a Case of Creeping Centralization. In C. Glendinning, M. Powell and K. Rummery (eds) *Partnerships, New Labour and the Governance of Welfare*, Cambridge, Polity, pp.167-182.

Deakin, N. and Edwards, J. (1993) *The Enterprise Culture and the Inner City*. Routledge, London.

Department of the Environment (1988) *Action for Cities*. London, HMSO.

Department of Transport, Local Government and the Regions (2001) *Urban Regeneration Companies – Learning the Lessons*. London, DTLR.

Desai, M. (1989) Is Thatcherism the Cure for the British Disease? In Green F. (ed) The Restructuring of the UK Economy. London, Harvester Wheatsheaf, pp.299-312.

Driver, S. and Martell, L. (1998) *New Labour*. Cambridge, Polity Press.

Driver, S. and Martell, L. (2002) *Blair's Britain*. Cambridge, Polity Press.

Etzioni, A. (1993) *The Spirit of Community*. New York, Crown.

Fairclough, N. (2000) *New Labour, New Language?* London, Routledge.

Gamble, A. (1994) *The Free Economy and the Strong State*, 2nd Edition. London, Macmillan.

Giddens, A. (1998) *The Third Way – The Renewal of Social Democracy*. Cambridge, Polity Press.

Green, F. (ed) (1989) *The Restructuring of the UK Economy*. London, Harvester Wheatsheaf.

Hastings, A. (1996) Unravelling the Process of 'Partnership' in Urban Regeneration Policy, *Urban Studies*, Vol.33, No.2, pp.253-268.

Haughton, G. and While, A. (1999) From Corporate City to Citizens *Urban Affairs Review*, 35(1), 3-23.

Healey, P., Davoudi, S., O'Toole, M., Tavsanoglu, S. and Usher, D. (1992) *Rebuilding the City : Property-led Urban Regeneration*. London, E&FN Spon.

HMG (1977) *Policy for the Inner Cities*. London, HMSO.

HMG (1998) *Bringing Britain Together : A National Strategy for Neighbourhood Renewal*. London, HMSO.

HMG (2000) *Our Towns and Cities : The Future – Delivering an Urban Renaissance*, Cmnd.4911. The Stationery Office, London.

Hill, D. (2000) *Urban Policy and Politics in Britain*. Basingstoke, Macmillan Press.

Home Office (2002) *Community Cohesion*. London, The Home Office.

Imrie, R. and Thomas, H. (eds) (1999) *British Urban Policy* (2nd ed). London, Sage.

Lawless, P. (1989) *Britain's Inner Cities*, 2nd edition. London, Paul Chapman Publishing.

Lawless, P. (1996) The Inner Cities : Towards a New Agenda. *Town Planning Review* 67(1), 21-43.

Lawson, N. (1992) *The View from No.11 : Memoirs of a Tory Radical*. London, Bantam.

Mackintosh, M. (1992) Partnership : Issues of Policy and Negotiation, *Local Economy* Vol.7, No.3, pp.210-224.

Mackintosh, M. and Wainwright, H. (eds) (1987) *A Taste of Power : The Politics of Local Economics.* London, Verso.

Mayo, M. (1997) Partnerships for Regeneration and Community Development. *Critical Social Policy,* 52, pp.3-24.

Mayo, M. and Anastacio, J. (1999) Welfare Models and Approaches to Empowerment : Competing Perspectives from Area Regeneration Programmes, *Policy Studies,* Vol.20, No.1, pp.5-21.

Neighbourhood Renewal Unit (2001) *New Deal for Communities.* London, DTLR.

Neighbourhood Renewal Unit (2002) *Changing Neighbourhoods, Changing Lives.* London, DTLR.

North, P., Valler, D. and Wood, A. (2001) Talking Business : An Actor-Centred Analysis of Business Agendas for Local Economic Development, *International Journal of Urban and Regional Research,* Vol.25, No.4, pp.830-846.

Oatley, N. (1995) Urban Regeneration. *Planning Practice and Research* Vol.10 (3/4), pp.261-210.

Oatley, N. (ed) (1998) *Cities, Economic Competition and Urban Policy.* London, Paul Chapman Publishing.

Oatley, N. (2000) New Labour's Approach to Age-Old Problem: Renewing and Revitalising Poor Neighbourhoods – The National Strategy for Neighbourhood Renewal, *Local Economy,* Vol.15, No.2, pp.86-97.

Peck, J. (1995) Moving and Shaking: Business Elites, State Localism and Urban Privatism, *Progress in Human Geography,* Vol.19, pp.16-46.

Peck, J. (1999) New Labourers? Making a New Deal for the 'Workless Class', *Environment and Planning C: Government and Policy,* Vol.17, pp.345-372.

Pollitt, C. (1990) *Managerialism in the Public Service.* Oxford, Basil Blackwell.

Raco, M. (1997) Business Associations and the Politics of Urban Renewal: The Case of the Lower Don Valley, Sheffield, *Urban Studies,* Vol.34, No.3, pp.383-402.

Schuftan, C. (1996) The Community Development Dilemma: What is Really Empowering? *Community Development Journal,* Vol.31, No.3, pp.260-264.

Stewart, M. (1998) Partnership, Leadership and Competition in Urban Policy. In N. Oatley (ed) *Cities, Economic Competition and Urban Policy,* London, Paul Chapman Publishing.

Stoker, G. (ed) (1999) *The New Management of British Local Governance.* Basingstoke, Macmillan.

Tomaney, J. (2002) New Labour and the Evolution of Regionalism in England. In J. Tomaney and J. Mawson (eds) *England – The state of the Regions,* Bristol : Policy Press, pp.25-44.

Urban Task Force (1999) *Towards an Urban Renaissance.* London, E&FN Spon.

Ward, K. (1997) Coalitions in Urban Regeneration: a Regime Approach. *Environment and Planning A,* 29, pp.1493-1506.

Chapter 5

London: The Mayor, Partnership and World City Business

Peter Newman and Andy Thornley

Introduction

Over the past 20 years the government of London has undergone a series of radical changes. Abolition of the Greater London Council (GLC) in 1986 raised concerns about the coordination of local governance fragmented among 33 London Boroughs, numerous unelected London-wide bodies and local regeneration agencies, and about the lack of democratic representation in the big decisions shaping London's future. In May 2000 a strategic scale of government was re-established. The new Greater London Authority (GLA) comprised an elected London assembly and, for the first time in London, an elected mayor. However, the problem of coordination remains as the new mayor and assembly add to the already complex institutional landscape with its overlapping layers of public and public-private agencies engaged in planning and managing the capital. Additionally, the proliferation of unelected or indirectly elected agencies continues to raise concerns about the quality of local democracy. The issue of coordination and the question of representation form the main themes of this chapter.

In the unitary British state it is national government that creates, abolishes and recreates local governments. Since the establishment of the first institutions of modern local government in the nineteenth century, relationships between local and national government in London have been controversial. Political differences lay behind the abolition of the GLC, which was itself created in the hope of guaranteeing Conservative Party control of the strategic scale of government. Conflict between the current London mayor and national government continues to be a troublesome relationship. As we shall see this conflict has had important impacts on the approach and policies adopted by the mayor.

Following an initial review of institutional change since abolition of the GLC, the chapter then examines in detail the development of the new institutions of the Greater London Authority (GLA). We set out the ways in which mayoral government works focusing, in particular, on relationships between the mayor and business interests. At the end of the 1980s, business leaders took a more visible role in policy debate. It was those businesses with obvious strategic interests in the capital – British Airways, Grand Metropolitan Hotels, National Westminster Bank, British Telecom – that took the lead. Since the 1980s business has promoted the idea of London as a world city.

The new mayor, Ken Livingstone, supports this policy priority. The discourse of London as a world city gives some sense of direction to fragmented institutions and helps the mayor organise his relationships both with business and with central government. We discuss the contemporary roles of business in government and the bias towards business in decision-making in the new institutions of London government.

Following our review of the new scale of mayoral government, the chapter reviews issues of coordination and representation at sub-regional scale, looking in particular at the complex relationships between local government, business and public–private partnerships in central London. We conclude by summarising the key themes emerging at these different scales and re-examine explanations of the strong orientation of policy towards London as a world city and to the needs of business in maintaining the city's competitive edge.

The Changing Institutional Landscape

The Greater London Council (itself a relatively new institution set up in 1965) was abolished in 1986 along with metropolitan scale governments in other large UK cities. Between 1986 and 2000 there was only one tier of local government in London.[1] The London Boroughs (the 32 Boroughs and the City Corporation that governs the central financial district) managed most services, with joint borough committees engaged in some London-wide activities, including strategic planning. This fragmented government was compounded by the growth of public–private agencies established by British urban policy experiments: City Challenge; the Single Regeneration Budget; the New Deal for Communities; etc. Large regeneration areas – the Thames Gateway that links London with neighbouring counties to the east, for example – are managed through overlapping public and public-private partnerships.

After 1986 leading businesses took a greater role in the government of the city and new private sector agencies were established at London–wide and local scales to add to the complexity of governance. Encouraged by the Conservative government, the business lobby London First was founded in 1992. London First produced strategies and promoted inward investment and at a London-wide scale business seemed to offer a coordinating role.

During the 1990s national government also attempted to coordinate strategic decision-making. A Minister for London and a cabinet sub-committee were established to keep the government's hands on 'The UK's number one asset' (Gummer 1996). Government reorganised its London wide activities into the Government Office for London (GOL) in 1994. In the 1990s London government could be seen as being pulled in two apparently contradictory directions, towards greater fragmentation and increasing centralisation (Newman and Thornley 1997, analyse this tension during the 1990s).

---

[1] The Conservative government established new 'Government Offices' in all English regions in 1994. The Government Offices brought together staff from the central Departments of Environment, Employment and Trade and Industry and were expected to coordinate the government's regional expenditure and act as the 'eyes and ears' of government in the regions.

By the time of the election of the Blair government in 1997, London presented a complex pattern of governance. In response to demands from within the Labour Party and in response to a vociferous campaign from the London Evening Standard and other lobbies, the Labour government proposed a new experiment in government – the direct election of a London-wide mayor. The Standard's regular polling put the issues of coordination and the lack of a democratic voice for Londoners at the top of its readers' concerns. The Conservative and now Labour governments had also voiced the concern that London might be falling behind other European cities, especially those such as Barcelona that seemed to have effective strategic institutions. Indeed the mayor of Barcelona accompanied Tony Blair at a public debate organised by the Evening Standard and the Architecture Foundation, which endorsed the idea of a mayor for London.

The mayor and Greater London Assembly were elected (on the same boundaries as the former GLC) in May 2000. The mayor's responsibilities include strategic planning, economic development and transport. The mayor is required to produce a large number of strategic plans, but has limited powers to coordinate action across London. The institutions of the GLA add to an already complex pattern of governance. The Government Office for London remains and the Boroughs retain their responsibilities. The Labour government's programme of modernising local governments has given the London boroughs new responsibilities for coordinating partners within their boundaries, within which typically a multiplicity of local partnerships manage neighbourhood regeneration. These layers of government create institutional complexity. The complex, overlapping partnerships tend to obscure the questions of accountability and representation that concerned Londoners in the 1990s. Reform of London government gave Londoners the 'voice' they demanded (72% voted in favour of the GLA and mayor in a referendum in 1998), but the mayor has only limited powers. The development of the role of mayor has also been coloured by the personality clash between mayor Ken Livingstone and the British Prime Minister, Tony Blair. Livingstone was not selected to run as the Labour Party candidate , choosing to stand as an independent candidate and gaining 58% of the vote. Splits in the London Labour Party resulted in the Conservatives winning most seats in the Assembly elections. Relations both between mayor and Assembly and between mayor and national government have been shaped by conflicts between personalities. The consequence of these divisions however has, as we shall see, had an impact on the way the mayor works and the priorities he has pursued.

The Greater London Authority

In this section we examine firstly the nature of the new institutions of mayor and assembly and examine how the mayor has involved business groups in the government of London. The final part contrasts relationships with business with the GLA's attempt to include a wide range of interests in debate about London-wide issues.

The New Approach

The new Labour government pledged itself to greater transparency in government, to tackle the issues of the proliferation of unaccountable *ad hoc* bodies, and to devolve governmental power. They also indicated that they would give greater emphasis to issues such as social exclusion and environmental sustainability. Policy co-ordination, or 'joined-up policy thinking', were also priorities. The office of mayor was conceived as having strong executive powers. Alongside the mayor an elected Assembly would have a scrutinising and checking role. It was hoped that a strong mayor would overcome the problem of a lack of political leadership in the capital, and that the electoral processes would introduce greater transparency and accountability into strategic decision-making. One of the major features of the new model was that it would be a streamlined authority. There was no intention of returning to the huge bureaucracy that was a feature of the old Greater London Council.

The Greater London Authority Act was passed by parliament in November 1999. The mayor had responsibilities at a strategic scale, with the lower tier of government (the 32 London Boroughs and the City Corporation) remaining but not duplicating the new institutions. The GLA took over the powers of some existing quangos, with some new powers devolved from central government (the SRB budget for example). Most of the executive powers of the GLA are vested in the mayor. Alongside the mayor's executive and the elected assembly are four 'functional bodies' with specific responsibilities – the Metropolitan Police Authority, Transport for London (TfL), the London Fire and Emergency Planning Authority and the London Development Agency (LDA). By far the most controversial of these has been TfL. Government has been reluctant to hand over responsibility for London's tube network to the mayor. The mayor formulates London-wide strategies, proposes a budget, co-ordinates all the different partners, and makes appointments to the functional bodies. The mayor also has to produce an annual progress report followed by a 'State of London' debate and face a twice-yearly 'People's Question Time'. The Assembly has twenty-five members of whom 14 are elected on an area or constituency basis and 11 on a London-wide basis. The role of the Assembly is to scrutinise the mayor's activity, make the appointments to the permanent executive, and also appoint some members of the statutory bodies. The mayor reports to the Assembly each month and answers members' questions. The mayor's proposals and budget are reported to the Assembly for endorsement. If the Assembly has the backing of two-thirds of its members it can request the mayor to make amendments to legislation. The Assembly also has the power to set up committees of investigation into topics of their choice and draw upon outside experts to provide advice and information.

Whilst the mayor has a strong list of executive powers, financial autonomy is restricted. The GLA took over the central government grants that previously went to the various transport operators, and these are paid to Transport for London. However, this money can only be used for transport purposes. The GLA also inherits the existing public spending in London on police, fire, economic development and regeneration. The important aspect of these funding arrangements is that central government still has a controlling influence and the mayor cannot switch funds between the different statutory bodies. Central government retains a reserve power to set a minimum level for the police budget. A very small amount of local tax income comes from a precept on the Boroughs. The GLA's own resources are limited with the most significant sources of revenue coming from powers to impose congestion charging and workplace parking fees. The mayor has only a few hundred staff compared to the ten thousand or so employed by the GLC. With a limited budget, powers and staff, the mayor has therefore to rely on influence, persuasion and the support of partners.

The mayor is required to draw up a series of strategies for London that deal with transport, economic development, air quality, waste management, culture, ambient noise, and bio-diversity. The mayor also produces a report on the state of London's environment every four years. An economic development strategy was published within the first few months after the mayor's election. The first draft of a cultural strategy only appeared in spring 2003. Bringing all of these strategies together is the Spatial Development Strategy (SDS) – now called the London Plan. Development of the London Plan follows a statutory timetable, and the important transport and economic development strategies preceded it.

The GLA is a distinctive type of citywide government. The powers of the London mayor are very limited when compared with other big city mayors. The mayor lacks the comprehensive powers and large staff (40,000) of the mayor of Paris, or the substantial budget ($37bn) of the mayor of New York. But the new institutions, with their apparently open style of government, do respond to the questions about representation and democracy that Londoners raised in the 1990s. However, as we shall see, the new public forums are perhaps not as significant as other aspects of mayoral government.

The Continued Dominance of the Private Sector

In the years immediately following abolition of the GLC, business interests had a visible influence on strategic policy. The City Corporation (the self governing financial district, whose institutions remained untouched by local government restructuring in the 20th century) took a more active promotional role and campaigned for citywide projects such as a millennium project in Greenwich. The business alliance London First, set up in 1992, had the ear of central government and established the question of London's international competitiveness at the centre of policy debate. London First produced a series of London wide strategies, including a transport strategy. Following the government's initiative in 1995 London First cooperated with the London boroughs to produce a 'London Pride' prospectus identifying policy priorities, with business competitiveness at the forefront.

The business lobby was generally supportive of the proposals for the GLA, although it wanted to ensure it had a voice in the new organisation. For example, London First argued that:

> 'London's prosperity and competitiveness depends on business. For London to remain competitive, business needs access to decision-making, a coherent voice to articulate its needs and the ability to make things happen. The GLA and its agencies must work in close concert with business.' (quoted in Kleinman 2001).

However, as Harding et al. (2000) suggest, business interests are not likely to seek influence in all aspects of strategic policy to the same degree. They suggest that the focus will be on local economic policy and land use planning. Certainly in the London case business interests focused a lot of attention on economic development strategy in the lead up to the establishment of the GLA. After the White Paper on the government of London had been published in 1998 various organisations with an interest in the London economy approached central government Ministers and offered their skills and resources to prepare the ground for the London Development Agency. This Agency would be under the responsibility of the Mayor and would take the lead in preparing the economic development strategy. This idea was accepted by government and the London Development Partnership (LDP) was set up with the aim to 'establish a business-led board' that would work to 'fill the strategic gap' in economic development thinking for London (LDP 1998a, p.2). Their first report, 'Preparing for the Mayor and the London Development Agency' (LDP 1998b), was produced by the end of 1998 and a draft economic development strategy published in January 2000 just in time to pass on to the new Mayor. The eventual economic development strategy, produced by the LDA at the end of 2000, drew heavily on this work. The LDP Board, as well as containing representatives of various public bodies, included representatives from the CBI London Region, London First, London Chamber of Commerce and Industry, and the Corporation of London. Many of these business representatives were to continue to sit on the LDA Board. The LDP Board itself had a great deal of overlap with its predecessor, the London Pride Partnership, and the issues it identified for priority treatment – business competitiveness, transport – were also similar. Meanwhile the London Chamber of Commerce and Industry (LCCI) had been developing its approach to the new London government. At the time the national government issued its consultation paper on the GLA in 1997, the LCCI commissioned a report from Ernst and Young (1997) on how business might best interact with the new London government. The report argued that the best Mayor would be a high profile person from the business community. This wish was not fulfilled, but the report made two other important recommendations, that 'private sector expertise should be deployed at the highest level in the GLA – in the Mayor's office – to help develop and implement strategies' and that 'for business to have an effective role in the GLA, it must be able to speak with a single voice' (p.iii). The LCCI took up this last point and promoted the idea of a London Business Board to provide a focus for business interests. The London branch of the Confederation of British Industries and London First agreed to join this Board and they held many meetings of business interests in the lead up to the Mayoral election, producing a document called 'The Business Manifesto for the Mayor and the GLA' that set out their priorities. It identified competitiveness as the key focus for the Mayor:

> 'The health and global competitiveness of London's economy must be at the heart of the GLA as the pre-requisite for achieving all other policy aims. All the GLA's policies must be tested against the aim of promoting a strong, stable, diverse, competitive, sustainable and flexible economy' (London Business Board, p.2).

Transport was identified as 'the Mayor's key challenge and business' top priority' (ibid, p.5).

The London Development Partnership established a group focusing on attracting inward investment and visitors to London, including the key promotional organisations, the London First Centre and the London Tourist Board. The group produced a report in May 2000, 'Promoting the World City: Memorandum for the Mayor and the GLA'. The purpose of the report was to ensure that the new London Authority would prioritise working with business to promote London and focus on improvements that would attract more inward investment. They presented the vision that by 2004 London would be 'The World City'. They said,

> 'When London First was set up in 1992, there was a belief in the business community that London was losing its status as a world class city. Over the last eight years, London has not just regained that status, it has reinforced its position as Europe's business capital and, along with New York and Tokyo, is one of three world cities. With the arrival of the directly elected Mayor and the LDA, the authors believe that London can aspire to be the undisputed World City by 2004'. (LTB and LFC 2000, p.9).

They looked to the mayor to prepare key development sites, improve transport, fund an International Convention Centre, improve air quality, skill levels and deal with crime prevention.

In the period from 1986 to 2000 business was active in shaping strategic priorities and in seeking to influence the shape of the new institutions of governance. The new mayor responded positively to the business agenda. The mayor's manifesto stated that Livingstone expected to work closely with the business community. He said that as mayor he would 'work with the Corporation of London and major City institutions to ensure London remains the financial capital of Europe', and 'support jobs and competitiveness in London by working with businesses and business organisation'. He said that he would only be able to succeed if he worked, 'with the active involvement of successful entrepreneurs and business people'. Given the mayor's limited powers and budget it is not surprising that he would seek out potential partners with resources to put behind his priorities. Such a stance is reinforced by the difficult relationship between the mayor and central government. Having left the Labour Party in order to run for office, Ken Livingstone could not expect to exert much influence over his former colleagues. Influential business leaders might be more friendly. One of the earliest acts of Livingstone's mayoralty was to set out how this working relationship with business would operate. The mayor prepared a framework for relations with the business community, drafted by management consultants KPMG. 'The Mayor and Relations with the Business Community' (GLA 2000) set out, in 26 pages, Livingstone's approach to working with business. The document states that:

> 'At the heart of the Mayor's job is making sure that London's success as a city economy continues. This requires more than just taking account of business issues in making decisions. It means forging an effective and productive partnership with business.' (GLA 2000, p.1)

It continues that in this partnership, 'the mayor intends to share his ideas and priorities with business so that a mutual relationship between the Mayor's office and business exists at an early and all subsequent stages of policy development.' (ibid, p.5). The document lists seven key features of the mayor–business relationships: openness: frankness; confidentiality; partnership; proactivity; reciprocation; and professionalism. It is worth quoting at length on the issue of confidentiality, as this was particularly important in moulding the nature of the business access in the agenda setting process. Under the heading of confidentiality the report states that:

> 'A precondition for open and frank discussion at an early stage in policy development is mutual confidentiality.   In the relevant areas the mayor will keep the details of his discussions with business confidential and will look to business to do so also.   Internally within the Mayor's office the Mayor will make clear to all staff that any breaches of confidentiality destroy the possibility of dialogue with the business community and that he will expect them to respect fully the need to maintain this confidentiality.' (ibid, p.6).

As soon as the mayor took up office his relations with the business community became a top priority. The London Business Board increased in importance after the establishment of the GLA because the mayor made it clear that he wanted a coherent view from business and the Board fitted into his idea of a Business Advisory Forum.

London's business organisations are well resourced and have long experience of engaging with local and national government. They developed contacts with all levels and departments of the GLA, not just with the mayor and his office, and this was something the mayor supported. For Livingstone, '[The mayor] also relies on business establishing strong relations with his Cabinet adviser on City and business, with senior officers of the mayor's office and GLA, and has given specific instruction that a substantial part of the time of relevant key officers must be given over to relations with outside business bodies.' (GLA 2000, ibid, p.18).   Business organisations dealt regularly with officers of Transport for London and the London Development Agency and the mayor appointed several business people to the boards of these two bodies, many of whom had also held office in the business organisations. Judith Mayhew, at the time political leader of the Corporation of London became the Mayor's City and Business Advisor. Very soon after his election the mayor set up bi-monthly meetings with the London Business Board. A representative of one business organisation said they were in touch with the mayor's office 'weekly if not daily.' Business groups were also pro-active in arranging meetings with the GLA; for example, London First set up a series of breakfast meetings for members on issues concerning the Strategic Development Strategy.

Business access to the agenda setting process can therefore be seen to have a number of significant characteristics. First it established an early presence in the process. Its activity before the mayor was elected, through its involvement in the LDP and the London Business Board, meant that it was well prepared to instil its priorities into the process. The Economic Development strategy was the first to be drafted, drawing on this earlier work, and was therefore in a good position to influence the other strategies. Michael Ward, the chief executive of the LDA responsible for the strategy, stated that this was the aim in getting the strategy done quickly. This early involvement of the business lobby meant that it was less necessary for them to engage in the later stages of general consultation or Assembly scrutiny. This focused approach meant that business access was established right into the centre of power through regular and confidential discussions.

Relationships between business and government in London suggest continuity over the years since abolition of the GLC. To the close business links with central government in the 1990s have been added close working with the mayor. The emphasis on London as a world city that developed in the 1980s continues into the mayor's economic and spatial development strategies. The ambitions of the government reforms in London for open and accessible government are brought into question. Let us briefly look at the how the mayor has developed relationships with other interest groups.

The 'Big Tent'

Many of the mayor's actions have suggested that he is taking a consensus-seeking and inclusionary approach – sometimes referred to as the 'Big Tent'. As we will see, he has been involved in a wide range of consultation processes.

In his manifesto, Livingstone promised to 'introduce the most open, accessible and inclusive style of government in the UK'. His election manifesto itself is characterised by embodying ideas from a range of interests. Livingstone's late emergence as an independent candidate meant he couldn't run on the Labour Party manifesto, but had to come up with his own. His team therefore relied heavily on input from outside bodies, particularly Friends of the Earth. The Green Party also says many of the ideas were originally theirs. He continued this inclusionary approach in his use of his power to make appointments to the various Boards within the GLA. These appointments came from a wide range of interest groups and political party allegiance.

Livingstone announced in autumn 2000 that he would undertake a process of 'stakeholder consultation' to help him formulate his priorities. Eighteen stakeholder groups were identified. The stakeholder groups varied in nature; some represented sections of society such as the elderly or young people, while others were organised groups like the Trade Unions or Academic Institutions. By July 2000 the Civic Forum, based at the London Voluntary Service Council, had a membership of 325 organisations, including voluntary and church bodies and minority organisations.

Within a few weeks of taking office, the mayor announced the formation of six 'policy commissions' covering housing, crime and community safety, environment, London health, equalities, and the Spatial Development Strategy (the London Plan). The commissions were to meet, debate, and make suggestions about policy directions for the GLA. These policy commissions, which were not provided for in the Act, were meant to represent the range of interests involved in each policy area. The criteria for including people in these commissions were unclear and varied between each one. This weakened the legitimacy of the commissions. However, the Spatial Development Strategy Policy Commission and the Environment Commission brought together an impressive range of experts and NGOs, with some business people as well. The SDS Commission had over 40 members divided into eight working groups. Meetings were not public. This commission had a strong business element. It also included architects, academics, developers and NGOs like the Pedestrians Association. Representation on the Environment Commission included the Black Environment

Network, Waste Watch, the London Tree Officers Association, as well as the National Health Service and London Electricity.

Although the SDS Commission did not produce a report, the one from the Environment Commission (GLA 2001) contained 100 recommendations. Although most of these were unexceptionable, there were a few controversial items: the commission said the mayor should focus on reducing the need to travel by encouraging local employment; and should oppose 'the relentless growth of airport capacity and air traffic.' Both of these points sat uneasily alongside the mayor's priority to retain London's position as a competitive world city. Although the commissions were told that their reports would form the basis for the mayor's agenda setting process, this proved not to be the case. The mayor was not going to deviate from the priorities in his manifesto and his officers had already done extensive preparatory work in some policy areas.

The new government institutions have brought new relationships in city wide policy making. But there is a clear difference between the open and consultative approach across a range of policy issues and the consistent and confidential relationship with business. In policy this is expressed through the priority for London's role as a world city. The world city separates sectors of the economy and the priorities clearly expressed by London First and other business lobbies. This has brought the mayor into conflict with other groups, through his support for more high buildings in the central area for example, and underwrites the approach to transport that hopes to secure new investment for rapid movement across central London.

The strong business lobby has been a consistent feature of London governance since the 1980s. The mayor's role, through leadership and partnership is to attempt to coordinate priorities at the strategic scale. But, as we have said, the mayor and GLA are just part of the layering of governance institutions in London. Maintaining London's world city role is also the objective of other scales of governance. In the next section we examine issues of coordination, partnership and business representation at a sub-regional scale, and the focus on institutional change in the central area of the capital.

Sub-Regional Partnership – Governing the Central Area

In this section we concentrate on the central area of the capital, in particular on the City of Westminster and the area around the financial district defined in policy documents as the 'city fringe'. The themes of institutional complexity and business influence are readily apparent at this sub regional scale. The first part looks at coordination across the central area. In the second part we look at institutional developments in Westminster and the city fringe and at the lead taken by business in the small-scale management of parts of the city.

The Central Area

The actual importance of 'global' functions to the London economy is debateable (Buck et al. 2002) but development pressure on the central area of London is distinctive. Whereas in other sub regions there may be more similarity with trends in other British cities, the central area is unique. Central London accommodates national government and international financial business services in addition to substantial cultural assets and its role as the UK's primary tourist destination. The Corporation of London has resisted institutional reform and continues to manage the financial services district. The City of London has traditionally competed with the City of Westminster with its government and ceremonial spaces. In the 1990s these two authorities came together with boroughs on the other side of the Thames. This Cross River Partnership (consisting now of 12 public and private partners) focused in particular on tourism and transport issues. Westminster and the City Corporation are also united in a wider sub-regional grouping – the Central Area Partnership.

The Central London Partnership (CLP) brings together the boroughs of Camden, Islington, Lambeth, Southwark, Kensington and Chelsea, Westminster and the Corporation of London. The CLP also includes a range of other public sector bodies including the Metropolitan Police and Transport for London and central area businesses. The CLP, which predates mayoral government, produced an Action Plan for Central London in 1998 and an Environment and Transport Strategy in 2001. Despite the arrival of the mayor, the CLP still sees the need to speak up for central London, which it regards as the, 'economic driver of the capital and south east region' (Central London Partnership 2003 p11). A revised Action Plan was published in 2003:

'Over this time, we have witnessed, and welcomed, the re-creation of a regional tier of government, with the first directly elected Mayor and his 'family' of regional agencies .... [the] CLP has been at the forefront of working with these new bodies..' (Baldry L 2003).

There is clearly no concern that the CLP may duplicate the mayor's functions. The CLP echoes the mayor's complaints about the transfer of resources from London to other regions and voices concerns about transport. A priority for the CLP is managing areas under pressure and the Partnership has led a campaign to adapt US models of managing business and tourist environments. The CLP leads the 'Circle Initiative' (a public-private partnership) that has been supporting prototype Business Improvement Districts (BIDs) in the areas of Waterloo, Bankside, Holborn, Paddington, and Piccadilly (Circle Initiative 2002). The central London boroughs have been attracted by the apparent success of BIDs in New York where local business pay enhanced property taxes, for instance to maintain clean streets and safe public spaces around stations and in tourist districts. The New York model developed out of the city's financial crisis in the 1970s and the desire to contain the general growth of local taxes. In New York BIDs have been controversial as BID security forces have moved street vendors and beggars out of their areas and BID managers clashed with the mayor. But for the CLP such disadvantages are outweighed by the strong attractions of clean streets and safety for visitors and employees of businesses in the designated areas. The CLP lobbied government for legislation for a British version of BIDs that would allow increased local business taxes in specified areas. The five experimental areas are managed by public-private boards and are anticipating national legislation in 2004. We can see that business-government cooperation is an important aspect of the governance of London at strategic scale and at the scale of the central area sub region.

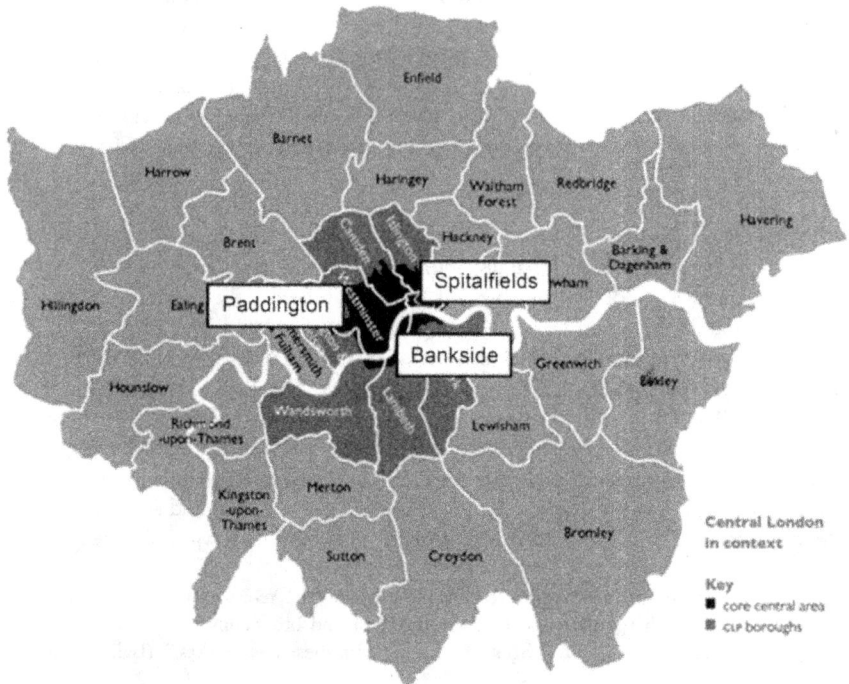

Figure 5.1 Central London: Proposals for Business Improvements Districts

Partnership at Borough and Neighbourhood Scale

The reform of government in London in 2000 retained a two tier structure. This type of structure is unusual among large 'world' cities – in both New York and Paris, for example, the lower tier enjoys few powers. But the government was keen to introduce mayoral government quickly and wider debate – about the appropriate boundary of Greater London, and about the role of the City and boroughs – would have created controversy. The City and the boroughs retain their functions. But the boroughs have also been undergoing significant change as part of the government's wider project of modernising local government. Some boroughs have elected their own mayors. All are involved with government's ideas about community governance and the coordination of services and regeneration across public sector and between public and private sectors.

The 2000 Local Government Act brought new responsibilities for community governance. In London's central area Westminster City Council manages the Westminster City Partnership as the 'local strategic partnership' coordinating public, voluntary, community and business sectors. The Partnership's 2002 Westminster City Plan is sponsored by the health service through the Primary Care Trust, the Metropolitan Police, voluntary and community sectors, local business, local regeneration partnerships, and residents groups. The Plan focuses on a nationally defined 'local' agenda of civic renewal, crime and disorder, and improving community health (Westminster City Partnership 2002). But the voice of central area business is as strong here as it is in setting the mayor's priorities. The conclusion from the City Plan's consultation on its investment programme is that, 'Westminster should continue to be a leading world city' (ibid, p.9). The Plan interprets this to mean dealing with infrastructure renewal, congestion and enhancing heritage and cultural assets. Westminster's potential BIDs, in Paddington and around Piccadilly in the West End, are focused on preventing crime and managing congestion. The Plan identifies sub areas with particular problems where again the strategy is partnership working and cross sector collaboration. From neighbourhood scale upwards, the governance of London is constructed from layers of overlapping partnerships and strategies. The world city orientation gives a common theme as partnerships and strategies proliferate.

Westminster is at the heart of the central area with its national and world city functions. Its deprived areas are managed through complex and overlapping partnerships. Around the City of London we find a similar pattern. The Cityside Regeneration Company in Spitalfields was built on a series of national government funding programmes starting with the Bethnal Green City Challenge (£7.2m) in the early 1990s, continuing through Single Regeneration Budget Round 3 (SRB3; £11.4m) in 1997, and most recently Single Regeneration Budget Round 5. The longevity of such regeneration programmes suggests that 'world city London' also has some intractable problems.

The SRB3 programme was titled 'Building Business' and substantial investments have been made in small businesses (Cityside Regeneration 2002). Projects attempted to link big business (banks on this edge of the City) and small business, and to exploit the tourism and entertainment potential of a distinctive ethnic neighbourhood through promoting local festivals, rebranding 'Banglatown' and its cuisine. Success in this project created the largest cluster of Bangladeshi and Indian restaurants in London. But large commercial developments, both in banking and in the culture industries, have escalated land and property values and excluded many from the benefits of regeneration. The Cityside Regeneration Company is a public and private partnership that includes an international bank and national construction companies. Through substantial new development projects, Spitalfields has been joined more firmly to the City. The regeneration projects benefit from expertise from big business partners, but the reasons for business involvement are straightforward. ABN Amro constructed a new headquarters at the end of the 1990s and local regeneration brings environmental improvements that make the area more attractive and safer for its employees (Clarke 2001).

In this part of the central area the world city agenda – providing new spaces for financial services and new tourist draws – is as visible as in Westminster. Local government and regeneration companies work through the often-complex mechanisms of partnership, bringing business into local decision making. These processes developed in the 1990s, and continued following the arrival of mayoral government. At neighbourhood scale government is being modernised. Overlapping partnerships and strategies add to the density of governance institutions.

Conclusions: Governance in the World City

In London, as in other British cities, the national agenda of modernising and regionalising government have substantial impacts. The experiment with mayoral governance and the GLA is unique and the new institutions have taken a distinctive path. The debate about the future of London government in the 1990s reflected strong views about democratic representation and calls for a 'voice for London'. The new institutions express many of the ambitions of the Blair government in opening up decision-making in accountable and community-oriented governance. Londoners have also got the leadership that was demanded. But in this chapter we have emphasised two main arguments. Firstly, the new institutions have not replaced other bodies and the governance of London continues to be characterised by complex layers of public and public-private agencies operating at different scales. The second argument highlights the dominance of business in setting a strategic direction for the city and the significant role of business in managing at sub regional and neighbourhood scales. The question of representation in policy making continues to be an issue. Neither the institutional complexity nor the role of business are new. We have pointed out the continuity in the ways in which the city has been governed over the years since the abolition of the Greater London Council.

At the neighbourhood scale there is enthusiasm for the idea of Business Improvement Districts. The newest local scale initiative is oriented to the needs of business and the management of safe and clean local environments to retain staff and encourage visitors. The BID pilot areas include the Holborn business district, the newly emerging business quarter in Paddington, the established tourist zone around Piccadilly, and the new tourist districts in Waterloo and Bankside. But many other areas want to try the BID experiment, including Spitalfields. The BID is therefore seen as the next source of regeneration and renewal funds for neighbourhoods and local areas.

The central area sub-region has a clear idea of its national and international role and of the need for a (cross party) public and private sector lobby to speak up for the heart of the city. Notwithstanding the arrival of mayoral government and a host of new London wide strategies, the Central London Partnership revised and relaunched its own central area strategy. Partnerships and strategies proliferate. For the central area, any coordination problem is overcome initially by a common focus on London world city.

The LDA published its strategy quickly in order to influence other strategies. The LDA manages regeneration and neighbourhood renewal budgets but the dominant part of the mayor's economic agenda is retaining London's world city status. This ambition continues the stance adopted by London First and others since end of 1980s. We noted the policy continuity from London First to the LDP and to the London Business Board, and the privileged position of business in the mayor's decision making. The Business Board made dealing with business interests straightforward.

The priority accorded to 'world city' reflects this representation but also the more complex relations between the mayor, business, and central government. A world city orientation suggests that London has a different vocation to other UK cities and therefore is not in direct competition with them for national resources. The mayor argues that London subsidises other regions, and that the businesses that create that subsidy need to be nurtured. If London has a different vocation then national investment decisions – Cross-Rail, an Olympic bid, etc. – have to be seen in a different light. The mayor has a group of business leaders who make the case for these resources and the case for London. Arguments for London may thus be treated more favourably by central government than they would be if seen as merely the demands of a maverick mayor. The world city orientation is good politics.

References

Baldry, L. (2003) Chairman's Welcome: *Action Plan for Central London,* Central London Partnership.

Buck, N. *et al.l* (2002) *Working Capital: Life and Labour in Contemporary London,* London, Routledge.

Central London Partnership (2003) *Action Plan for Central London,* Central London Partnership.

Circle Initiative (2002) *The Circle Initiative,* www.londonbids.info.

Cityside Regeneration Ltd (2002) *Cityside SRB3 Final Report,* Cityside Regeneration Ltd.

Clarke, A. (2001) *Regeneration and Competitiveness: The London Finance Industry"s motives in regeneration,* MA Thesis, University of Westminster.

Ernst and Young (1997) *How Regional Government Works with Business,* London, Ernst and Young.

GLA (2000) *The Mayor and Relations with the Business Community,* Greater London Authority Mayor's Office.

GLA (2001) *Environment and Sustainability: A GLA Report from the Mayor"s Policy Commission the Environment,* Greater London Authority Mayor's Office.

Gummer, J. (1996) *Celebrate London's success — don"t knock it.* Talk to the Evening Standard/Architectural Foundation debate on the Future of London, January.

Harding, A., Wilks-Heeg, S. and Hutchins, M. (2000) Business, Government and the Business of Government, *Urban Studies,* Vol.37, n.5/6, p.pp.975-994.

Kleinman, M. (2001) *The Business Sector and the Governance of London,* paper presented to the European Urban Research Association Conference, Paris.

LDP (1998a) *LDP Bulletin,* London Development Partnership.

LDP (1998b) *Preparing for the Mayor and the London Development Agency,* London Development Partnership.

London Business Board (undated) *The Business Manifesto for the Mayor and the Greater London Authority,* London Business Board.

London Tourist Board/London First Centre (2000) *Promoting the World City: A Memorandum for the Mayor and the GLA,* London, LTB.

Newman, P. and Thornley A. (1997) Fragmentation and Centralisation in the Governance of London: Influencing the Urban Policy and Planning Agenda, *Urban Studies,* Vol.34, no.7, p.pp.967-988.

Westminster City Partnership (2002) *Westminster City Plan,* WCP.

Chapter 6

The Contradictions of Partnership: Sheffield From Steel to Urban Regeneration

Philip Booth

For more than twenty years in the United Kingdom, we have been accustomed to the fact that urban regeneration must of necessity involve actors other than the local authorities and central government. As a result, the concept of partnership has become generalised to the point at which it has lost much of its heuristic value. Moreover, because the practice has become so widespread, there is the danger of imagining that it amounts to very much the same sort of thing wherever it is met with. But half the charm of comparisons derives from the way in which numerous studies attest to significant differences between places as to how urban policy is put into effect.

Before turning to the case of Sheffield, three preliminary observations need to be made. We are used to pointing to the divide between London and its metropolitan area and the cities of provincial England. But there exist equally strong differences between the provincial cities, even where they are not particularly far apart. So Sheffield has little in common with Birmingham (Gaetano and Lawless 1999), is equally different from its neighbour, Manchester, particularly in terms of its collective identity (Taylor et al. 1996) and its rivalry with the other big conurbation in Yorkshire, Leeds, is part of local folklore. It is necessary, therefore, to reflect on the history and geography of Sheffield in order to contextualise partnership as part of local authority practice.

The second observation concerns the concept of partnership itself which is often profoundly ambiguous. Much has been written about public-private partnership and the opening up of local decision making on urban policy to the private sector, not only in promoting economic recovery but also in actually making and implementing policy. Such an approach was characteristic of all the Conservative administrations of the 1980s. This was of course hardly surprising, given that the involvement of the private sector was part of the ideology of the party. The Conservatives wanted to ensure that the market was given as free a rein as possible to develop the nation's wealth and to become involved in policy making that had hitherto been entirely the preserve of public bodies. But the concept of partnership as it was expressed at the outset of Margaret Thatcher's premiership undoubtedly went through an evolution. Even before the accession to power of New Labour, the theme of partnership had long since moved beyond the simple involvement by private sector employers in economic policy and had been applied to the social as well as the economic development of deprived areas. Local residents' groups began to find themselves becoming part of the partnerships that were to address these problems.

The third observation has to do with the links between actors. To talk about partnership between the private sector, local associations and local authorities says nothing about the power relations represented by the links between actors. As we shall see, at Sheffield the private sector chose to work with the city council from 1986 onwards. But using the term 'private sector' suggests a coherent group with clear organisational boundaries. Is this really the case? And if not, what exactly does this 'private sector' amount to? Which private actors were in fact involved and in what ways? What effects did their involvement have? If these are the kind of questions that are raised in talking of the private sector, they are equally relevant to the public realm. Certainly, it is possible to identify certain general tendencies in local government in Britain, but each of the major cities has its particular characteristics which relate to the specific conditions that obtain there, and which distinguish one city from the next. Imagining that the 'private sector' and the 'public sector' are homogeneous phenomena and the relations that have with one another are stable over time is to ignore both the richness and the subtlety of the process of partnership.

These preliminary remarks suggest the need first of all to sketch out the social, industrial and political context of Sheffield. This is followed by a presentation of the forms of local governance that have structured public policy in the town in the long term. Finally, the chapter turns to an analysis of partnership within Sheffield, taking a historical perspective to consider, among other things, relations between government departments and between central and local government.

The Character of Sheffield

Sheffield is one of the seven large conurbations of northern England. Nevertheless, it is town which has the fewest attributes of a metropolitan area, either in its physical or in its socio-political characteristics. It has long been described as the biggest village in England, a description that relates as much to its location as to its industrial base.

At the outset, nothing suggested that Sheffield would become a major industrial centre. Its geographical position was hardly of the most encouraging. It was no more than a small settlement set in a forest clearing next to fords over the rivers Don and Sheaf which were protected by a small fort. Road, and later rail, links to the outside world were difficult and until the 19th century, the town made little impact at the national level. Its one comparative advantage, which was apparent by the end of the middle ages, and which was to become the basis of its developing economy after the industrial revolution, was its tradition as a centre for the production of cutlery. The existence of iron ore in the area, the abundance of wood for the charcoal needed in forging blades and above all the water power to turn the gritstone wheels on which the blades were honed, all assured the quality of Sheffield knives.

To begin with, cutlery production was a rural industry, which depended for the watercourses for its location. But from the end of the 18th century the cutlers began to collect in the town itself, a movement confirmed by the arrival of superior ores from Sweden and the use of coal rather than charcoal for heating forges. In this first industrialisation, Sheffield remained dominated by cutlery manufacturing, which by now was located around the town centre and just upstream on the banks of the River Don (Wray et al. 1993).

The industrialisation of the cutlery industry began to generate increasing activity in research into new ways of making steel: the discovery of ways to produce steel of very high quality and in large quantities became the source of Sheffield's second industrial era. From the middle of the 19th century, steel-making became the major industry in Sheffield and located on the flat land available in the city, to the north-east of the city centre in the Lower Don Valley. The industry gradually spread over some four miles up to – and indeed beyond – the boundary of Sheffield with neighbouring Rotherham. This industrial expansion led to a rapid growth of the population: the town of 30,000 people at the beginning of the 19th century had become a conurbation of half a million by the eve of the First World War.

There are two points to make about this history of Sheffield. First, the industrial character of the city had a profound effect on its social character. Cutlery manufacturing was – and until relatively recently remained – dominated by small firms run by bosses who were known in the local dialect as the 'little mesters', who displayed an independence of mind and were close to their workers. The large steel-producing firms, in spite of their size continued this strongly established tradition and this was reflected in the social relationships within firms. It explains the relative absence of confrontations between workers and their bosses and the deeply rooted tradition of mutual dependence. It is also the case that Sheffield was characterised by the weakness of its middle class as compared with Manchester or Birmingham. Secondly, Sheffield for long remained isolated from the rest of the country. This isolation was both a product of its geographical location but also from the way in which the local economy was dominated by cutlery and steel. If each of the major conurbations had its own characteristic industry, all with the exception of Sheffield had fairly soon diversified their manufacturing base. Sheffield was in competition with to the west, Manchester, which effectively became the capital of northern England and to the north, by Leeds, which developed as a regional centre within Yorkshire for banks and insurance companies.

Local Governance in Sheffield

Sheffield, somewhat isolated from the rest of the country, solidly working class, and heavily dependent on the steel industry, developed during the course of the 19th century a tradition of mutual dependence between local actors that persisted in the 20th century. From 1926 to the present day, the city council has been controlled by the Labour party with only two brief interludes totalling no more than three years in total (Seyd 1990). Until the 1970s, the council intervened in a paternalistic way to provide the local population with a growing number of services of which housing and education were the most important. The period between 1926 and 1939 saw the creation of a series of local authority housing estates, typically in the form of semi-detached houses with gardens. After the Second World War, during which bombing badly damaged the city centre, the city threw itself into the process of clearance of slums that had not been demolished before the war. For the reconstruction of these areas the city looked to Le Corbusier's Radiant City model, most notably present in the scheme for Park Hill, which gained world-wide celebrity at the time.

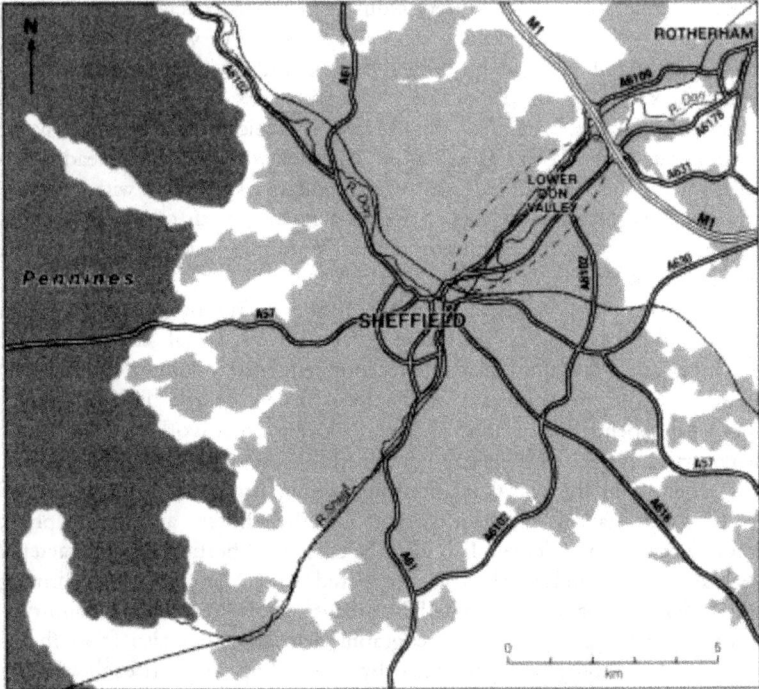

Figure 6.1 Sheffield

This period of benevolent paternalism, which lasted until the 1970s, was based on a consensus among local actors. In the first place, the council's legitimacy resided both in universal suffrage and in the close ties between local councillors, many of whom were workers in the steel industry, and the unions. Secondly, local fortunes were closely tied to the steel and engineering industries. For as long as these industries remained strong, the consensus could endure. It was the collapse of the steel industry at the end of the 1970s that led to a radical change in the classic political relationships between social groups.

From Municipal Socialism to Public-Private Partnership

The first signs of economic difficulties tied to the global over-production of steel and competition with the developing world began to appear in the 1960s. To begin with, Sheffield was protected from this global recession in the steel industry by virtue of the fact that its steel industry was oriented towards the production of high quality steels, not quantity. But Sheffield's over-specialisation did not protect it from the general crisis and in the end it was as badly affected as other steel-producing cities in Western Europe and North America. A reduction in the work-force was already beginning during the 1970s, but the full force of the recession did not hit Sheffield until the early 1980s. In Sheffield, the crisis was all the more brutal for having arrived late, and for the fact that for so long the city considered itself immune from the difficulties faced elsewhere. In 1970, Sheffield and Rotherham together had some 60,000 employees in steel and related industries; which had dropped to 16,000 in 1987 and 10,000 by 1993 (Tweedale 1993). For an urban area of some 750,000 people, such a loss of jobs marked the start of a crisis without precedent. The effects on Sheffield's political system were to be of two kinds. On the one hand, the city council began to look for ways in which to combat unemployment and restructure the local economy. On the other hand, the crisis profoundly altered the way in which the council worked, both by changing the composition of the council itself and modifying its relations with the outside world.

Three phases in this process can be identified. The first was characterised by an interventionist municipal socialism that became heavily implicated in economic development. The second phase was structured around attempts by the city council to involve the private sector, while at the same time retaining a leadership role. The third phase, which has persisted to the present day, has been characterised by a new type of partnership, in which there has been a reconfiguration of local leadership in policy making for urban regeneration with new actors involved, and the rise of local residents' groups. Each of these phases represents the attempt by the city council to overcome local difficulties. But it is also undeniable that what happened in Sheffield runs in parallel to the changes in national policy set by successive governments.

Municipal Interventionism 1980-1986

Until the 1970s, the local political elite in Sheffield consisted of people who were heavily identified with the city's industrial base. But the reduction in the number of employees in the steel industry led to a change in the social composition of the council, with the arrival of councillors who were often employed in public sector services rather than in industry and who were typically to the left of the Labour party. This radicalisation of the council was matched nationally, which was marked by the capture of the councils of Liverpool, Manchester and Greater London by the hard left. There was, however, one distinct difference between these places and Sheffield. Where in most of the country the shift to the left took place abruptly and gave rise to a hostile press campaign, in Sheffield the transition was more gradual with a gentle increase of the newer type of elected representative. This transformation was only completed when in 1980 the leadership of the council passed to David Blunkett, who was to become Home Secretary.

In 1980, the council for the first time began to take on board as part of its policy framework the question of economic development, which hitherto had been left to industry. The new leadership believed that if the local economy was to be revived, the council would need to be involved in the creation of jobs and that administrative effort should be devoted to that end. In 1981, the council created the Department of

Employment and Economic Development (DEED) whose objectives were to reduce unemployment, to assist in the creation of jobs but also to promote the social and community economy. The first director of DEED was a radical economist whose ideas accorded well with the political orientation of the council (Seyd 1993).

It should be noted that the underlying philosophy of DEED was far from one of the involvement of the private sector in urban regeneration. According to its director, DEED's objectives were to 'liberate the resources of the local state and put them at the service of the working class movement, the women's movement and community based movements' (Benington, cited by Lawless 1990, p.140). To achieve this end, DEED initiated a certain number of small projects that aimed at creating 'socially useful' products. Under its leadership, a concert hall for the profoundly deaf and sanitary facilities for the physically disabled were produced (Blunkett and Green 1983). The council also increased the number of its own employees and developed innovative ways of funding cooperative ventures. But, as Lawless (1990) noted, resources were far from being adequate to deal with problem of restructuring the economy in the early 1980s and the radical socialism of the City Council dissuaded potential investors from outside the city.

From Municipal Interventionism to Partnership 1986-1997

The change to strategic thinking within the City Council that was to characterise the second half of the 1980s was brutal and caused considerable bitterness among local leaders. The change was the result of external pressures as much as of the crisis in the local economy. The Thatcher government, known for its mistrust of local authorities, and above all of those which, like Sheffield, were controlled by the Labour Party, sought ways of limiting their power. Two constraints in particular had a fundamental impact on the ability of local authorities to act and their room for manoeuvre. One was the ceiling placed on the ability to raise local rates. The other was the imposition of the Urban Development Corporations which were charged with the renewal of the country's most badly affected industrial areas.

The limits imposed by central government on the ability to determine its own rates were at the heart of the crisis within the council and which 'encouraged' Sheffield to move towards partnership with the private sector. The Rates Act 1984 required government to calculate the total expenditure of each local authority in the light of local need, and then to impose a ceiling on the amount of money that could be raised locally through the rates. In 1985 the City Council was faced with an unpalatable choice: either to reduce its expenditure or to refuse to set the legal maximum rate, and so risk a political stand-off with central government and challenge through the courts. After heated internal debate, the City Council decided to avoid confrontation and reduced its budget. It was clear that from now on, any policy drive to restructure the local economy would have to be carried out in partnership with the private sector.

In 1986, the council set up its first partnership organisation, the Sheffield Economic Regeneration Committee (SERC) which brought together representatives of local industry, the unions and local associations. Its mission was to be a place for discussion and to produce a strategic vision for the future of Sheffield that would engage all sectors. SERC was set up by the City Council, but it is clear that the private sector, as represented by the Chamber of Commerce and Industry, was already conscious of the risks of isolation. In the early 1980s, local industry and the City Council had tended to blame each other for the economic crisis, but realising that this did nothing to advance the cause of economic restructuring nor to promote the image of Sheffield in the outside world, the first steps towards informal contact had already been taken even before the creation of SERC, with the aim of ensuring that at least in public the private sector and the City Council spoke with a single voice (Lawless 1990).

In order to explain this apparent convergence of interest between the council and local actors represented by the Chamber of Commerce and Industry, analysts at the time made much reference to north American theories of urban regimes and growth coalitions. But Strange (1996, 1997) and Lawless (1990) have shown the limits of the applicability of such theories to the British context, and particularly to the case of Sheffield. First of all, SERC was a creature of the City Council and can be seen as a product of 'strong' local government which does not obtain in the United States. Then, too, SERC differed from other such structures in England in its limited membership. There were no representatives of the media, of central government or of multi-national enterprise. It remained essentially a local affair.

The impact of SERC was disappointing. A great deal of effort was made to promote Sheffield's image, and these efforts were without doubt genuinely cooperative. But the idea of a strong private sector promoting a strategy for the development of the city in partnership with elected representatives never took off. The Chamber of Commerce and Industry was as devoid of innovatory ideas as the City Council. Nevertheless, one of SERC's initiatives did have significant effects. In 1987, it commissioned the consultants Coopers and Lybrand to report on the future of Sheffield. The report recommended the creation of a flagship development that would give the city the kind of renown that had formerly come from the steel industry.

The occasion for realising that ambition came with the decision in 1988 to host the 1991 World Student Games. To do so meant creating a number of important sporting facilities that would in turn give Sheffield a profile as a centre of sporting excellence. Others have described in detail how this project unfolded (Seyd 1993; Foley 1991) and for the purposes of this chapter we need only concern ourselves with certain aspects. First, we should note that if the initiative was enthusiastically taken up by elected representatives as an antidote to the gloom of the preceding decade, local private enterprise was no less enthusiastic. Then, too, there was no evaluation done of the real impact of the project on the local economy. And then again, it rapidly became clear that the necessary sponsorship from local business and from the media was not going to be forthcoming, and that the City Council was going to have to take financial responsibility for a much larger proportion of the cost than had been anticipated. As a result, the City Council is still to this day paying off debts incurred. Finally, there was criticism that the infrastructure created did not best serve the needs of local residents.

The results of this first period of partnership were fairly disastrous, therefore. Nevertheless, the memory of the games in June 1991 and the cultural festival that accompanied them, remains fixed in the local consciousness. And indeed, the games marked the beginning of a series of projects which were to transform the physical character of the city. In economic terms, the World Student Games were highly controversial. Their positive effect on the morale of the city's inhabitants, who had been facing economic crisis was nevertheless considerable.

Sheffield Development Corporation and the Regeneration of the Lower Don Valley

By setting up SERC, the City Council wanted to demonstrate clearly that it placed great importance on the regeneration of the Lower Don Valley in partnership with local business interests. However, central government had other ideas, and preferred to impose the structural solution already applied in London Docklands, in Liverpool and elsewhere by creating an Urban Development Corporation that would take over from the City Council responsibility for economic development, land management and development control within defined boundaries. The corporation's board was made up of members nominated by government, not elected, and the private sector was well represented. Three places were reserved for local elected representatives, but the chairman was a well known local businessman, not a local councillor. In this way, it followed a model that was very different from that advocated by the City Council itself. Indeed, the logic of the Sheffield Development Corporation (SDC) showed itself to be very different from that of SERC: its primary task was to raise land values and to create infrastructure in order to encourage real estate investors to develop, on the understanding that development of land would of itself generate employment (Dabinett and Ramsden 1999).

The City Council's view of this imposition was something of a paradox. If it regretted the fact that its own approach to partnership had not been recognised by central government and resented the size of the area that had been placed in the hands of the SDC, it did not, unlike local authorities within London Docklands, refuse to cooperate with SDC. From the beginning, two Labour councillors occupied the places reserved for them on the board, and the city's planning department provided an agency service for development control within the SDC's area, a task which ensured a degree of oversight of SDC's activities.

In its early years, SDC had to face the fall in land values, which affected the country as a whole at the end of the 1980s, but which had strong local consequences for Sheffield. An initial plan for the area proved far too ambitious and had to be rethought. But by the end of its existence in 1997 – like all the Urban Development Corporations, SDC was conceived as a fixed-term agency – the transformation of the Lower Don Valley was an accomplished fact, even if some of the achievements which SDC claimed had been initiated by others. This was particularly the case with Meadowhall regional shopping centre, whose origins lie in an agreement between the City Council and a private promoter before the SDC had been created.

In the ten years that divide the creation of SERC and the winding up of the SDC, local governance in Sheffield had undergone a major transformation.

- The City Council had moved away from its radical position of the 1980s by accepting that partnership with the private sector was unavoidable in the economic development of the city. This new division of work meant that it would no longer act alone in this field. At the beginning with the creation of SERC, the town hall had nevertheless attempted to control partnership: council departments serviced the committee and its chair was a councillor. With the setting up of SDC, the City Council saw clear diminution of its power to control, even if the council's strategy to minimise the impact of SDC was to cooperate, and not to oppose.

- Participation by the private sector in urban regeneration nevertheless remained limited. Both in SERC and in the SDC representatives of business were entirely local, which for certain commentators explains the parish-pump outlook of both organisations. Conflict within business leadership was also a limiting factor. The SDC, with its chairman recruited from local industry nevertheless found itself in conflict with small businesses in the Lower Don Valley who were faced with the compulsory purchase of their premises for the creation of a new road (Raco 1997).

- The model offered by the SDC, that of a single-purpose government agency, separated from the City Council, was to become the preferred solution to the problem of urban regeneration after 1997. What began as an ad-hoc structure became the dominant mode of intervention.

- The partnership at the heart of the SDC focused on physical renewal, not on economic development through job creation. It was from these beginnings that the idea of physical project planning became more and more current in the thinking of local decision makers and certainly explains the importance accorded to the World Student Games and to major building projects thereafter. Nevertheless, it is very had to evaluate the impact of the games and of the major infrastructure development undertaken by the SDC on employment. On the other hand, in the context of the generalised social and economic crisis within Sheffield, it is difficult not to give credit for such development, which, if it did not actually spell the end of economic crisis, at least bore witness to a significant urban dynamism.

Towards Fragmented Local Governance 1997-2003

1997 marked both the end of the SDC and the accession to power of a Labour government under Tony Blair. In the field of urban regeneration, the change of government was marked as much by the continuity of previous policy as by radical change. Indeed, even before the end of the SDC and the beginning of the Labour administration, national policy had undergone an important evolution. This evolution owes its origins to the realisation that the results of the Urban Development Corporations had been unsatisfactory and that there had been confrontation between the UDCs and the directly elected local councils.

In 1991, the Conservatives had established the City Challenge programme, destined to integrate separate financial initiatives into a single package for urban regeneration in given areas. Local authorities were invited to make bids for funding under the double heading of economic and social development and physical development. The allocation of funding depended on the quality of the bid and the nature of the partnership with the private sector and local associations. The experience with City Challenge was sufficiently encouraging for the government to expand the programme as the Single Regeneration Budget (SRB) from 1995 onwards. Sheffield did not receive funding under the two rounds of City Challenge, but government did grant the city a substantial tranche of funding in the first phase of SRB for the north-west sector of the city. The SRB programme was renewed four times under the Conservatives and twice under the Labour government after 1997. Sheffield received funding under each phase such that almost all of its most deprived areas were covered.

SRB was welcomed by locally elected representatives because it appeared to give local authorities back the initiative in urban regeneration. However, local authorities were not in sole charge because the management of SRB areas had to be given to partnership boards on which the private enterprise and neighbourhood associations and other members of civil society were represented. Sheffield found it necessary to establish an agency – the Sheffield Regeneration Agency – to coordinate the activities of local SRB boards. It is also the case that certain areas developed their own agencies as a counterweight to the City Council.

In Sheffield, SRB became the occasion for the council to transfer some of its responsibilities to the neighbourhoods, in a sort of internal decentralisation. The partnership that the government was looking for needed a mobilisation of the local population through local neighbourhood committees; in Sheffield these had already come into existence to fill the gap that the City Council's withdrawal from social development had left. Inevitably, these local committees took on responsibility for the preparation of SRB bids and to help them, some areas, notably Manor, set up technical services of their own. Those areas that lacked such technical support found the process hard to manage. SRB had important successes, but processing bids for funding and accounting for expenditure when funds had been granted was a complex task.

The Decentralisation of Powers: the Creation of the Sheffield First Partnership

SRB led to a diffusion of local administrative capacity to the neighbourhoods, even if the neighbourhoods did not necessarily have the capacity needed. But SRB was far from being the only form of association between the local authority and other interests in this third phase of partnership, which in fact was witness to an explosion of consultative bodies, some created before 1997 and others a direct result of national policy afterwards. The most important of these was to be the Sheffield First Partnership.

SERC had never really become the place in which a strategy for the city as a whole could be worked out. Excluded from the Lower Don Valley by the creation of the SDC, SERC's membership was also too large to be effective. In 1992, the City Council set up a new partnership body, the City Liaison Group. This group whose membership was far smaller than that of SERC, was intended to represent the five key sectors in the city's economy: the City Council, higher education, private enterprise, the health service and the development agencies (primarily in the first instance SDC). Its task was to give leadership in the field of local development. According to Dabinett and Ramsden (1999), the success of Sheffield in obtaining SRB funding was very largely due to the group.

Indeed, the City Liaison Group proved to be far more effective than SERC in developing an urban regeneration strategy for the city, but at the price of a certain loss of control for the City Council. For while the Leader of the Council chaired the group and the Chief Executive was on the board, the group was not a City Council committee in the way SERC had been, and it acted to some extent independently of the town hall. Moreover, tasks that had traditionally been handled directly by the council itself were now being transferred to the group.

This independence of action was confirmed after 1997. Public-private partnership became key to Labour thinking about urban regeneration and the delivery of local services following the election. The government took on board a certain number of Conservative policies such as the Private Finance Initiative, which gave private enterprise an important place in public services. However, the Labour government was also attached to the idea of greater community involvement, by participation as much as by representation, and this thinking informed New Labour reforms. The City Liaison Group corresponded to thinking at national level and from 1998 its role was reinforced under its new guise as the Sheffield First Partnership, which henceforward was to have its own technical services whose director had been a civil servant and not a former town hall employee.

The confirmation of this new role for the former City Liaison Group was as much a product of the arrival of a new Chief Executive, with a remit to restructure administration within the town hall, as it was of national policy. The legitimacy of the Sheffield First Partnership relied on the continued role of the Leader as its chair and the presence of the Chief Executive on its board. But more than that, it proceeded to hold a series of meetings in which a wider public was encouraged to voice its opinion on the future of the city. At the same time, the partnership's field of activity was widened by the creation of a series of sub-partnerships with specific remits, each of which was finally accountable to the board of the Sheffield First Partnership as a whole. All this could be seen as the way in the City Council was once again trying to maintain control over urban regeneration by innovating in the creation of partnership structures. But it was as much a sign of the weakening of the City Council as of its strength. The indebtedness of the city following the World Student Games had led to a decline in the Labour vote, such that it lost overall control of the council between 2000 and 2002. The Sheffield First Partnership assured a continuity which overcame the loss of Labour's majority in the council itself.

National policy aimed revitalising local democracy resulted in the Local Government Act 2000 which required local authorities to develop a Community Development Strategy whose preparation was to be the responsibility of what was to be called a Local Strategic Partnership. The Sheffield First Partnership 'naturally' fulfilled this role and was duly recognised by central government in 2002 (Sheffield First 2002). The importance of this recognition was by no means simply symbolic. Aimed particularly at the most deprived neighbourhoods, the Community

Development Strategy was intended as a replacement for SRB as a vehicle for channelling funds to where they were most needed.

Partnership for Urban Renewal

This remodelling of local administration went hand-in-hand with other tendencies in national policy for urban regeneration. One aspect of Labour's programme was to promote the renewal of local democracy. Another was the renewed emphasis on the architectural quality of urban renewal projects which was set out in the report of the Urban Task Force, *Towards an Urban Renaissance*, which had been led by the architect Richard Rogers (Department of the Environment, Transport and the Regions 1999). One of the recommendations of the report, which was taken up immediately by central government, was the setting up of Urban Regeneration Companies (URC) whose purpose was to deal with the problems of urban regeneration in a coherent and effective way.

The model for these new structures was partly that of the UDCs of the 1980s. As with UDCs, URCs were intended to be lightweight, single-purpose organisations, which would involve private sector enterprise in urban renewal and which would aim for high quality environmental design. Once again, urban regeneration was seen in terms of physical projects and would be property-led. The URCs were not to have the same structure, however. They were to be companies limited by guarantee whose shareholders would be the local authority, the Regional Development Agency, English Partnerships and representatives of the local private sector. Their accountability was, therefore, through company law.

Central Sheffield became one of the first three areas to acquire a URC in 2000, under the title of Sheffield One. Chaired by the chairman of Barclays Bank, its objective is the transformation of the commercial heart of the city. In less than a year, it had produced a master plan for the centre and is currently engaged in negotiations with developers on several major projects in partnership. If all these projects are implemented, the centre of Sheffield will have undergone a radical transformation (Sheffield One 2002). It must be emphasised that Sheffield One has not acted alone and has drawn upon the strategy already set out by the Sheffield First Partnership. Moreover, insofar as the City Council is one of the shareholders and representatives of the private sector are already involved in partnership with the council, Sheffield One represents as much a continuing trend as a profound shift. It is more in its manner of operation that the change is to be noted, and particularly the fact that its legitimacy is guaranteed by company law rather by the traditional mechanisms of local government. If Sheffield One has mobilised an array of actors concerned with urban regeneration, it remains relatively distant from the population at large.

Conclusion

The history of partnership and participation in public policy making in Sheffield over the past twenty years represents in microcosm the changes that have been happening nationally in local governance in Britain, even if this evolution is in part modified by local conditions within Sheffield.

Certainly, this evolution suggests a profound transformation in local authorities, in the urban policy they have developed and in the way their activities are legitimated. We can say that in Sheffield this shift is all the more dramatic for the fact of its long tradition of paternalistic municipal socialism pursued by the Labour council, and the depth of the crisis suffered by the city in the 1980s. There has been a transition from a traditional, representative democracy, legitimated by universal suffrage, to a participative democracy, in which the City Council acts in partnership with a series of actors who are increasingly given powers of deciding on matters of public policy.

Part, at least, of this transformation was due to New Labour's political platform, with its desire to reinvigorate local democracy by encouraging community involvement, even if the way in which that term is used at national level has been vague. On the other hand, this rediscovered interest in the notion of community takes us back to the cooperative movement of the 19th century, and has often been invoked by the left (Blunkett and Green 1984), even if in its simple form it is ill-adapted to the present day needs of multicultural and pluralistic society.

The case of Sheffield demonstrates how complex the change has been. The City Council has clearly changed its functions over the course of the last 20 years. Until the end of the 1970s the City Council was the major provider of services to the city's residents. From the 1980s, it became involved in fighting unemployment and in job-creation, and eventually in the process of urban regeneration, relying more and more on partners outside the town hall. Such an evolution was itself not a simple one. The partnership that became increasingly central to public policy making in the city was of divergent types, and the organisational structures to support did not necessarily relate well one to another. Nevertheless it is possible to identify three separate strands in the ways of establishing partnership.

- The model of the benign paternalism that marked municipal socialism before 1980 had already undergone a major change with the arrival in power of the radical left, with its ambition to influence economic development. The force of events led the City Council to realise that it could not hope to act without the cooperation of the private sector, and specifically the Chamber of Commerce and Industry. However, in this first phase, partnership remained centred on the town hall, for the creation of SERC was a way of mobilising the private sector without the loss of municipal control. SERC was more a mechanism for consultation than for full participation.
- Partnership as understood by the Conservative government was of a quite different order. In the name of efficiency, the Conservatives made the involvement of the private sector an essential component of local governance, in an attempt to marginalise democratically elected institutions. At Sheffield, the SDC, just as with the other Urban Development Corporations, was directed by a board whose members were nominated by central government, not elected. The board's accountability for their wide-ranging powers was nominally to Parliament, but in practice that control was attenuated.
- The third phase of this history in Sheffield is the most complex. In it, multiple forms of partnership become apparent, once again in part a response to national policy intentions. On the one hand, the Labour government was genuine in its

desire to mobilise civil society through the Local Strategic Partnerships and the Community Development Strategies. On the other hand, they maintained forms of partnership that originated with the Conservatives. The use of single-purpose, lightweight organisations, which would be effective in promoting urban regeneration has continued, with all the consequent doubts about accountability and public involvement. The invention of Urban Regeneration Companies was part of this process, which has now been further confirmed by the re-introduction of a UDC to tackle urban problems in the West Midlands (English Partnerships 2003).

In Sheffield, the City Council has attempted to limit the centrifugal forces that the multiplication of agencies has given rise to by maintaining leadership of Sheffield First and Sheffield One. But it is still open to question whether these new forms of governance have in fact weakened municipal control and whether a genuinely participative democracy has emerged.

References

Blunkett, D. and Green, G. (1984) *Building from the Bottom: the Sheffield Experience.* Fabian Society Tract 191, London, Fabian Society.

Dabinett, G. and Ramsden, P. (1999) Urban Policy in Sheffield : Regeneration, Partnerships and People. In R. Imrie and H. Thomas (eds) *British Urban Policy: an Evaluation of the Urban Development Corporations*, London, Sage, pp.168-185.

Department of the Environment, Transport and the Regions (1999) *Towards an Urban Renaissance.* Report of the Urban Task Force, London, Spon.

Di Gaetano, A. and Lawless, P. (1999) Urban Governance and Industrial Decline: Governing Structures and Political Agendas in Birmingham and Sheffield, England, and Detroit, Michigan, 1980-1997, *Urban Affairs Review*, Vol.34, pp.546-577.

English Partnerships (2003) *Annual Report.* London: English Partnerships.

Foley, P. (1991) The Impact of Major Events: a Case Study of the World Student Games and Sheffield, *Environment and Planning C*, Vol.9, pp.65-78.

Lawless, P. (1990) Regeneration in Sheffield: From Radical Intervention to Partnership. In M. Parkinson and D. Judd (eds) *Leadership and Urban Regeneration*, London, Sage, pp.133-151.

Raco, M. (1997) Business Associations and the Politics of Urban Renewal : the Case of the Lower Don Valley, Sheffield, *Urban Studies*, Vol.34, No.3, pp.383-402.

Seyd, P. (1990) Radical Sheffield : From Socialism to Entrepreneurialism, *Political Economy*, Vol.38, pp.335-344.

Seyd, P. (1993) The Political Management of Decline 1973-1993. In C. Binfield *et al.* (eds) *The History of the City of Sheffield 1843-1993*, Vol.1, *Politics*, pp.151-185.

Sheffield First (2002) *Sheffield City Strategy*, Sheffield, Sheffield First.

Sheffield One (2002) *Reshaping Sheffield City Centre*, Sheffield, Sheffield One.

Strange, I, (1996) Participating in Partnership : Business Leaders and Economic Regeneration in Sheffield, *Local Economy*, August, pp.143-157.

Strange, I. (1997) Directing the Show? Business Leaders, Local Partnership, and Economic Regeneration in Sheffield, *Environment and Planning C*, Vol.15, pp.1-17.

Taylor, I., Evans, K. and Fraser, P. (1996) *A Tale of Two Cities: a Study in Manchester and Sheffield*, London, Routledge.

Taylor, S. (1993) The Industrial Structure of the Sheffield Cutlery Trades 1870-1914. In C. Binfield *et al.* (eds), *The History of the City of Sheffield 1843-1993*, Vol.2, *Society*, pp.174-210.

Tweedale, G. (1993) The Business and Technology of Sheffield Steelmaking. In C. Binfield *et al.* (eds) *The History of the City of Sheffield 1843-1993,* Vol.2, *Society,* pp.142-193.

Wray, N., Hawkins, B. and Colum, G. (2001) *One Great Workshop: the Buildings of the Sheffield Metal Trades,* London, English Heritage.

Chapter 7

Toronto's Reformist Regime, Municipal Amalgamation and Participatory Democracy

Julie-Anne Boudreau

Toronto is a typical North American city, with a proud skyline where Canada's financial sector thrives, and an ethnically diverse population of about 5 million people. Yet Toronto has also been known as a model city where local democracy flourishes. When, in 1997, the conservative provincial government of Ontario imposed the amalgamation of the city's six lower-tier municipalities and the upper-tier Metro Toronto, this participatory culture that had developed since the rise to power of reformists in 1972 was threatened. As of 1972, reformists ensured that the city centre remained vibrant by countering modernist development based on freeway construction and suburban growth (Lorimer 1970; Sewell 1972; Harris 1987; Caulfield 1988b; Sewell 1993; Caulfield 1994; Allen 1997).

The reformist coalition (it never was a political party) coalesced into an urban regime of a kind, opposing economic development if it impeded on the environment or social and cultural quality of the city.[1] Between 1962 and 1973, the number of office spaces downtown Toronto doubled, exercising enormous pressure on the central residential neighbourhood.[2] Reformists won most Council seats in 1972 by favouring instead a 'human scale' planning philosophy, focusing on downtown residential vitality, community development, public transit, public spaces, density, and diversity. This resistance to the pro-growth planning philosophy, prevalent in most North American cities at the time, resulted in the preservation of vibrant central

---

[1] In the Canadian parliamentary system, each governmental level functions with its own political parties, which are not structurally linked to those of other levels of government as it is the case in France and Great Britain. Hence, the Ontario Liberal Party is independent from the Federal Liberal Party. The municipal level is generally non-partisan as a result of the first reform movement of the turn of the 20th century that sought to eliminate corruption by depoliticising local politics. Citizens vote for independent municipal councilors in their respective ridings (councilors sometimes display affinities with provincial or federal parties, but these are not formalised). An important exception to this rule is Montreal, where the municipal electoral system is partisan. However, municipal parties are unique to Montreal and are not linked with provincial and federal parties.
[2] Between 1970 and 1980, the number of office spaces rose by 78 per cent, and again by 71 per cent between 1980 and 1993 (Lemon 1996, p.274; Filion 2000, p.173).

residential neighbourhoods, similar to that of the Greenwich Village celebrated by Jane Jacobs (Jacobs 1961).

Beyond the built environment, downtown Toronto has developed a participatory culture, first in the field of planning, then in virtually all decision-making sectors in the City structure. This participatory culture was not restricted to local democratic mechanisms implemented by the former City of Toronto. It also translated into left-of-centre political behaviour at all levels of government. At the local level, reformists created numerous citizen commissions, while also decentralising municipal offices towards less central neighbourhoods. Before amalgamation in 1997, the City of Toronto sought citizen input on more than 130 committees, in addition to the numerous public meetings, task forces, public consultations, and so on (City of Toronto 1997, pp.28-29). The City was well-known for its special programs on HIV prevention, domestic violence, day-care subsidies, drug abuse help centres, struggles to alleviate poverty and homelessness, recreation programs, the Toronto Arts Council, etc. This participatory culture contrasted with the political climate in the suburbs, which were known for supporting the provincial conservative government. Provincial elections show a pattern of conservative votes in the suburbs, and left-left votes in the inner city (New Democratic Party and Liberal Party).

This political cleavage between a centre-left City of Toronto and a conservative province of Ontario has prevailed since the end of the Second World War, with the exception of an interlude from 1985-1995. When the Tories came back to power in 1995, they positioned themselves much more on the right than in previous decades. With Mike Harris as their leader, the Tories adopted a neo-liberal turn that contrasted with the conservative tradition of Ontario. It is no coincidence that the official name of the Tories is the Progressive Conservative Party (PC) as it emphasised the statist and interventionist stance of the party, also known as Red Toryism. But in 1995, the PC was elected on the promise of a 'Common Sense Revolution' that transformed the party into a neo-liberal stronghold. A discourse on 'reducing the size of the state', 'reducing the number of politicians', 'cutting waste', and 'downsizing bureaucracy', led to a promise to cut taxes by 30 per cent while bringing the deficit to zero. In order to deliver, Harris engaged a series of social and education cuts. This affected the province's relation with municipalities given that they are partly responsible for delivering and financing these services. Following numerous policies of disentanglement, the Tories came to amalgamation.

This 1997-98 institutional and territorial policy profoundly changed the socio-political climate in Toronto. Many feared the weakening of the reformist participatory regime. These transformations, it is important to note, not only paved the way for the rise of a new neo-liberal socio-political elite, whose eyes are turned towards growth and international competitiveness, but it also opened the door for leftist critiques of the reformist regime to be heard. During the reformist years (1972-1998), critiques of the regime were certainly voiced loudly on the right wing of the political spectrum. But under the veil of a socio-political consensus about Toronto's 'unique' participatory and caring culture, critiques voiced on the left wing of the political spectrum were hardly heard.

This chapter begins with an analysis of the gap between Toronto's participatory culture and the new institutions inherited with the 1998 amalgamation policies. This explains the force of resistance to amalgamation. The following section presents three projects implemented by the new amalgamated City of Toronto in order to recreate the reformist participatory culture: the new City Plan, the Diversity Advocate, and participation in the municipal budget. The new conditions for local democracy are evaluated by detailing both right-wing and left-wing critiques of reformism, in order to highlight some of the problems with the still prevailing sense that Toronto is a model for participatory democracy.

GREATER TORONTO AREA AND THE NEW CITY OF TORONTO

Figure 7.1 Greater Toronto Area and the New City of Toronto

Angst Towards the Megacity: Amalgamation Clashes with an Entrenched Participatory Culture

In December 1996, the Harris government announced its forced amalgamation plans for Toronto.[3] The urbanised core of the GTA (formerly known as Metro Toronto) was to be amalgamated to form one giant municipality of 2.5 million. The rest of the GTA, known as the '905 area' in reference to their different telephone area code, was to remain untouched by the amalgamation reforms.

As soon as this amalgamation project became public, citizens in downtown Toronto mobilised. John Sewell, former reformist mayor in the 1970s called a public meeting where Citizens for Local Democracy (C4LD) was formed to struggle against amalgamation. A ferocious battle began (Boudreau 2000). It lasted until April 1997, when the government was able to have its amalgamation bill (Bill 103) passed in the provincial legislature. The new City of Toronto became reality on 1 January 1998.

C4LD was a non-partisan coalition serving as a clearing house for citizens opposed to amalgamation, most of them residing in central neighbourhoods and having a long history of participation in reformist municipal institutions. Of all political hues, these residents agreed on two core characteristics of the central city's participatory culture: the importance of local democracy and of a dynamic neighbourhood life supported by a dense and mixed built environment. As a movement, C4LD quickly became very visible in the media and mobilised thousands of residents for weekly meetings and various actions (demonstrations, municipal plebiscites, lobbying provincial opposition parties to organise a filibuster in the provincial legislature, letter campaigns, massive participation in public hearings on amalgamation, legal challenge to Bill 103, etc.).

Three main arguments were put forward by C4LD in opposing amalgamation: 1) a bigger city will dilute the power of downtown reformists in a council dominated by a suburban and 'less' democratic culture, while distancing elected representatives from their base given the bigger councillor: citizens ratio; 2) despite the province's constitutional power in municipal affairs, imposing amalgamation was considered illegitimate because it was not announced in the Tories electoral campaign, nor were any public consultations conducted before the fait accompli; and 3) combined with cuts in social services and education, amalgamation threatened local democracy by favouring the development of a neo-liberal local regime that would support the province's agenda.

It quickly became evident that the Harris government had underestimated the strength of Toronto's reformist participatory culture, a culture residents intended to preserve. Within a few years, it is true that with its 44 elected councillors, the new city

---

[3] The Greater Toronto Area (GTA) is composed of three geographical scales: 1) the GTA comprises the new booming suburbs for a total population of 4,68 million (approx. 41 per cent of the population of the province of Ontario and 15 per cent of the Canadian population) on 4,400 km2; 2) the new amalgamated City of Toronto (often called the Megacity), which was formerly organised around the upper-tier Metro Toronto and six lower-tier municipalities, for a total population of 2,48 million on 632 km2; and 3) the former City of Toronto (now amalgamated with its postwar suburbs and the upper-tier Metro Toronto Council), which had a population of 654,000.

has a ratio of councillors to citizens higher than before (Table 7.1). However, it remains difficult to assess whether such ratio actually make councillors less accessible. Nevertheless, Torontonians now have to deal with a more centralised decision-making structure and a more complex bureaucracy (the new city employs close to 50,000 people). After much debate, the new city created community councils, but in contrast with Montreal's new borough system (established in 2002), Toronto's community councils are not composed of councillors elected to serve specifically on a community council. They are composed of representatives elected to City Council and do not have a political staff of their own. They only have a power of recommendation in the fields of local planning, local transportation, local recreation services, and neighbourhood affairs. They are accountable directly to Council and to the mayor, just like other permanent committees on Council (these are: Policy and Finance, Administration, Planning and Transportation, Economic Development and Parks, Works, Community Services, Community Councils). Community councils' boundaries do not correspond to the former local municipalities and do not foster resident allegiance.

Table 7.1   Comparison of the Ratio of Elected Representatives to Citizens Before and After Amalgamation, Metro Toronto, 1996, 2001

	Elected to Metro Toronto Council	Elected at the local level	Total number of municipally elected representatives	Ratio representative: population
(Chair)	1	-	1	-
East York	1	9	10	1: 10,782
Etobicoke	4	13	17	1: 19,336
North York	7	15	22	1: 26,802
Scarborough	6	15	21	1: 26,617
Toronto	8	17	25	1: 26,149
York	2	9	11	1: 13,321
Total in 1996	**29**	**78**	**107**	-
City of Toronto after amalgamation (2001)	-	45	45	1: 55,144

Source: Adapted from Milroy 2000.

Two questions arise when considering the state of local democracy since amalgamation in Toronto:

1. Was the reformist regime displaced by the incorporation of suburban interests in the municipal machine as C4LD predicted? Can Toronto still position itself as a model of local democracy as it was the case during the reformist years (1972-1998)?
2. Is it justified to label the new urban regime as neo-liberal?

As it will become clear in what follows, the answer to the first question is no, as a new wave reformist mayor has been elected in the November 2003 elections after an interlude of five years under a pro-growth 'suburban' mayor. However, I suggest that despite the persistence of the reformist regime in Toronto, the answer to the second question is yes. Indeed, reformism in Toronto evolved preserving more of its original conservative elements than its more radical wing, and this in a context of generalised neo-liberalism. The following section explores the first question by assessing three of the new local democratic mechanisms implemented since 1998 and analyzing the public discourse on local democracy in Toronto. The second question is discussed by looking at critical views on reformism generally and on the practices of the new City specifically.

New Mechanisms for Local Democracy

Bill 103 had left open the question of the new city's organisational structure. Consequently, as soon as it became effective in January 1998, the new Council engaged a debate on the new City's structure, particularly with regard to the question of decentralisation. As mentioned above, the decision to create community councils with only recommendation power disappointed many residents who had hope to recreate the former municipalities through these community councils. However, the new City responded to resident pressure by creating other participatory mechanisms.

The New Plan: A Visionary Exercise

In April 1999, the City of Toronto launched an expert forum in order to create a planning vision for the new city. It was broadcasted on television. In June 1999, the City organised six public consultations and then mobilised hundreds of 'visionaries' such as politicians, community leaders, bureaucrats, developers, merchants, planners, university presidents, newspaper chief editors, media leaders, union leaders, and so on, in order to produce a 'vision'. The resulting new city plan establishes the City's priorities:

1. Maintaining an edge in a competitive market
2. Urban design
3. Reducing dependence on cars
4. Respect for the environment
5. Directing reinvestment and preserving neighbourhoods
   (City of Toronto 1999)

It states that 'globalisation and new communication technologies mean that jobs and investment can flow into our City with lightning speed – or flow out of it just as quickly'; that population growth is exciting, but it also exercises pressures on the environment, on housing capacity, and on transportation (City of Toronto 2000, p.2). The focus is on renewal, reinvestment, and public-private partnerships. In this visionary exercise, the new city attempted to seize the opportunity created by amalgamation to forge a new image centred on 'participation', 'urbanity', and competitiveness. This created a slippage from *open* participation to the *selection* of visionaries:

> Over the past 50 years, Toronto has been a real success story in North America. The question is, are we ready to take the next step forward and blossom as one of the great cities of the world? Do we have the energy and the will, the vision and the plan to capture the spirit of the 21st century? (City of Toronto 2000, p.1).

This visionary exercise was combined with Toronto's bid for the 2008 Olympic Games (organised by TO-Bid, a private entity).[4] The intention was to use the Olympics to revitalise the waterfront by injecting $17 billion of investment. Superficially, the combination of these three initiatives aiming to revitalise downtown Toronto recalls reformist-planning principles (local democracy and thriving central neighbourhoods). However, the result is a much greater centralisation of planning decisions. As Kipfer and Keil indicate, the visionary process leading to the new city plan was controlled by a handful of planners and consultants; the Toronto 2008 Olympic bid was organised by a private corporation, and the waterfront regeneration schemes were controlled by Bay Street (the Canadian equivalent of Wall Street) financiers (Kipfer and Keil 2002). Indeed, behind the gloss of an ecological, urban (rather than suburban), and participatory discourse, the City's new planning practices are in fact much less democratic than during the heyday of reformism.

Promoting Diversity in Local Participation

Partly in response to citizen pressures, partly as a legitimating strategy to instrumentalise the reformist 'positive' reputation, the new Council established, in March 1998, a Task Force on Community Access and Equity mandated to organise public consultations and to make recommendations on how the new Council could 'ensure the voices of the City's diverse communities continue to be heard' (Task Force on Community Access and Equity 1998). The mandate was to reinforce civil society, eliminate barriers to participation, reinforce community input into decision-making, and to continue to be a model employer with a civil service reflecting the city's diversity (Task Force on Community Access and Equity 2000). The Task Force's final recommendation included providing support for community organisation, to a proactive role in convincing private employers and other levels of government to eliminate discrimination in employment practices (from affirmative action policies to protection against sexual harassment, from the translation of municipal documents into several languages to the support of ethnic media coverage of city hall affairs).

On 15 December 1999, the Council adopted all recommendations proposed by the Task Force, creating five Community Advisory Committees (not to be confused with Community Councils): aboriginal affairs; access for disabled persons; status of women, ethnic and racial relations; and homosexual, lesbian, bisexual, and transgender affairs. These advisory committees established a formal mechanism through which Council can seek community advice on a given policy. They formalise certain procedures of the former City of Toronto. For many activists, however, these

---

[4] Toronto ultimately lost the race for the 2008 games.

advisory committees are a meagre compromise from their original demand to establish a permanent committee on Council for equity issues as well as an Equity Commission within the bureaucracy (Community Social Planning Council of Toronto 2000, p.1).

In December 2000, the City nominated its first Diversity Advocate (an elected representative responsible for diversity issues), who is mandated to coordinate all diversity-related activities with Council and Community Advisory Committees. The Diversity Advocate is further responsible for raising awareness on equity issues among private employers and non-municipal institutions. In nominating a councillor to this function, Council officially declared that Toronto is working towards the elimination of 'violence, racism, homophobia, homelessness, hate crimes, hunger, illiteracy and all barriers to human rights', recognising that 'the City of Toronto is increasingly becoming known as a city of diversity and that this very diversity of Toronto creates unique challenges for Toronto' (Toronto City Council Policy and Finance Committee 2001).

Hence, in its preparatory report for the City's participation to the United Nations' World Conference against Racism (UN-WCAR) held in Durban, South Africa in August 2001, the City of Toronto synthesised its anti-racist initiatives as follows (City of Toronto 2001):

1. Nomination of a Diversity Advocate
2. Adoption of an Action plan on access, equity, and human rights, based on the recommendations of the Task Force on Community Access and Equity
3. Creation of five Community Advisory Committees
4. Creation of task forces on linguistic equity and illiteracy, immigration and refugees, the elimination of hate crimes, equity in salaries
5. Adoption of a policy of non-discrimination
6. Affirmation of the necessity to adopt such policy for all municipal subcontractors
7. Adoption of a policy against workplace sexual harassment
8. Adoption of a policy on the elimination of hate crimes
9. Adoption of an employment policy
10. Support for access and equity programs
11. Response to the Federal government proposals for reforming Canadian immigration laws
12. Support for diverse awareness-raising and education programs against intolerance

When Toronto delegates came back from Durban at the end of the UN-WCAR conference, the City of Toronto adopted an action plan for the elimination of racism and discrimination, its main objective being to ease the participation of all residents in civic, economic, social, cultural, political, and recreational life in the city (City of Toronto 2002).

A Participatory Budget?

Since the beginning of the 1990s, the Metro Network for Social Justice (MNSJ) struggled to open the budgetary process to more public input (Conway 2004). After a decade of cutbacks in social and community services, and three years after amalgamation, the new City faced a serious budgetary crisis in 2001 (a shortfall of $300 million). The immediate reaction was to further cut social and community services, while increasing public transit fees and property taxes. Citizens quickly mobilised under the banner of Save Our City. In a coalition with unions, they created the Toronto Civic Action Network (TorontoCAN) in 2002. TorontoCAN's main objective was to open the budget process to the public in order to ensure that community services are not always the first to be cut. During the 2002 budget process, TorontoCAN organised a rally claiming the return of municipal subsidies to homeless shelters.

Responding to these pressures, the City published a community guide to help citizens participate in the budget process for 2003 (City of Toronto 2003). This is an interactive document translating technical budget language and offering calculus pages for citizens to establish their priorities. The guide had to be filled out by citizens and then mailed back to the City. Some public consultations were also organised. However, the City had no institutionalised obligation to take this citizen input into account. The process quickly discouraged activists who had in mind a participatory process more akin to the Porto Alegre experience.

Elected in November 2003, the new reformist mayor, David Miller, has pushed the participatory process a little further by organising citizen assemblies to discuss the 2004 budget. Called Listening to Toronto, the process was successful, but again, the budget committee did not have a series of concrete citizen proposals to follow (City of Toronto 2004).

Frustrations with the Reformist Regime: Practices, Discourses, and Critiques

A simple survey of documents produced by the City of Toronto highlights how the discourse on local democracy remains important (for a more detailed discourse analysis, see Boudreau 2003). On the surface, this seems to indicate that a central element of the reformist regime (citizen participation) is still very much alive in Toronto even after amalgamation.

However, there is an important gap between discourse and reality. This leads to the conclusion that Toronto's reformist regime, although still formerly in power under Mayor Miller, has partly slipped into an urban neo-liberal regime legitimised with numerous references to local democracy. Can we attribute reformist difficulties to amalgamation and budget cuts characteristic of the 1990s? C4LD had predicted the end of reformism with the amalgamation of the centre city with its post-war suburbs. It is true that this restructuring exacerbated difficulties; however, the reformist regime was losing steam before amalgamation. One could trace these tensions back to the class position of reformism and to the initial exclusion of more radical elements from the movement in the 1970s and 1980s.

From its beginnings, the reform movement was torn between a conservative faction preoccupied with the preservation of central neighbourhoods, green spaces, and property values, and a more radical faction whose main objective was to make more affordable housing available and enhance social justice. During the 1980s, academics (many of them involved in the reform movement) debated the social transformation capacities of reformism. Influenced by Castells' distinction between urban movement with no real transformative aim, and urban *social* movement aiming at structural changes (Castells 1983), intellectuals in Toronto wondered whether reformism was simply aiming at bettering the quality of life in downtown neighbourhoods, or whether it had broader transformative capacities (Harris 1987; Caulfield 1988a; Caulfield 1988b; Harris 1988). In other words, the debate questioned the importance of reformist middle-class positioning.

Almost two decades later, this debate still haunts reformists. More conservative reformists ultimately succeeded in marginalising the more radical elements of the movement that sought to mobilise beyond downtown middle-class residents. This social conservatism certainly helped to counterbalance the increasingly right-wing conservatism of provincial governments, but it failed to mobilise many Toronto residents for whom reformism was not responding to their social justice goals.

This tension was also present within C4LD. As a movement resisting amalgamation and neo-liberal cuts, C4LD certainly mobilised thousands of residents and dominated the political agenda for months. But most C4LDers were downtown residents, of middle-class background, with a university degree, and working in sectors heavily targeted by provincial cuts (civil servants, teachers, municipal employees, artists, media professionals, etc.) (Boudreau 1999; Boudreau 2000). Former reformist mayor and leader of C4LD, John Sewell, never hesitated to link the urbanity (defined as density and centrality) of downtown Toronto with the reformist political culture.

Residents in the pre-1950 city tend to forget this lesson about form. They assume it is normal to support programs which reduce social inequality, which assume wide participation in policy-making, and which allow for a large public realm, but those are simply values which are inherent to and amplified by a city form which includes significant levels of intensity, an active street life, and casual meetings of friends in public spaces. These social values are generated by the specific way the city is built; they do not seep out of the 'well-planned' city if it has low density and uses that are widely separated from each other (Sewell 2000, p.69).

This class positioning is linked to assumptions about what is urban, which implies the marginalisation of more radical claims to the city coming from immigrants, anti-poverty movements, homeless movements and squeegee kids.[5] This has exposed C4LD and reformists to leftist critiques (stemming out of the central city as well as the suburbs). Reflecting on C4LD, Keil synthesises very well this leftist discomfort with reformism:

> The middle class discourse of citizenship and local democracy has its own characteristic blind spots: social justice and identity difference, and, to a certain degree environmental justice. It remains a homeowner-based movement of small urban freeholders: it has not made the leap to an urban social movement capable of changing the meaning of the city (Keil 1998, p.161).

This leftist critique of C4LD is extended to the new local democracy mechanisms implemented since amalgamation. Many activists were rapidly disillusioned with the

---

[5] Squeegee kids form a youth movement based on a social critique and an alternative culture. Many squeegee kids are also squatters and/or live on the street. They work as windshield washers for a little money on street corners.

language of local democracy adopted by the new Council. They were suspicious since the beginning of the new city, but amalgamation could have provided an opportunity to radicalise reformism. However, as Kipfer indicates:

> The 'equity agenda' in the new Toronto is a largely symbolic affair, defined as it is by fashionable declarations of support for the 'diversity' and 'uniqueness' of Toronto as an immigrant city. Indeed, predominant notions of diversity and multiculturalism have become tools to promote the competitiveness of the global city of business. Multiculturalism is reduced to colourful festivals and tasteful food while the meaning of diversity is limited to a lifestyle choice and a marketing strategy to sell Toronto's Olympic bid and capture the youth market in designer clothing (Kipfer 1999-2000, p.17).

These critiques were formulated in reaction to amalgamation and budget cutbacks, but they also stem out of major socio-economic transformations. The population of Greater Toronto Area has considerably diversified since 1996. Many newcomers settle in downtown Toronto, but as well in the suburbs. Combined with processes of economic restructuring which tend to favour suburbs and with the slow economic revival following the harsh recession of the beginning of the 1990s, these transformations have opened the door to an aggressive discourse on the global competitiveness of the GTA. This is certainly not unique to Toronto, but the fact that this discourse on competitiveness goes hand in hand with a discourse on local democracy enshrined in a reformist participatory culture traditionally opposed to pro-growth development creates an awkward marriage. Toronto is thus undergoing an intense re-appropriation of reformist language (diversity, rights to the city, local democracy) in order to position the city on global markets by selling Toronto's quality of life, urbanity, and diversity.

The city's motto, 'Diversity, Our Strength', is seen by many as a marketing strategy, suspicion confirmed by the new official plan and the promotional discourse elaborated for the Olympic bid. Many question this symbolic celebration of multiculturalism as promoted by P.E. Trudeau's 1971 national policy of multiculturalism. As Croucher argues, this symbolic celebration of multiculturalism has erased the need for really understanding the material conditions of minorities: 'the Canadian state, by embracing and managing multiculturalism, has, in effect, co-opted political space available to minority groups for mobilisation or resistance along ethnic and racial lines' (Croucher 1997, p.335). Perhaps to make a difference on racism and inequalities, it would make more sense to think of diversity by asking: 'how and why a particular image or set of perceptions – in this case, ethnic and racial harmony – come to dominate the public mind and how, in the process, other conditions or complaints are not defined as problems or are denied a position on the polity's public agenda (Croucher 1997, pp.328-329). Studies on the level of poverty for non-whites in Toronto are alarming, and the diversity policies implemented by the new city are direct responses to these studies (Ornstein 2000). But many remain suspicious of these mechanisms celebrating diversity and imposing the illusion of racial harmony while silencing real debates on racism and discrimination. As Croucher indicated in 1997, Canadians are very uncomfortable with racial issues; they prefer to believe that the official multicultural policy is sufficient to counter racism (Croucher 1997).

Critiques of reformism also emerged from the right. They are much more widely heard than those stemming from the left. These right-wing critiques certainly played a central role in the provincial government's decision to amalgamate Metro Toronto in 1998. One of the immediate consequences of amalgamation was the dilution of reformist forces into a Council dominated by suburban interests and led by a pro-growth suburban mayor: Mel Lastman was elected twice between 1998 and 2003. Former mayor of North York, Lastman was pro-development, conservative and populist. Between 1998 and 2003, conservative-leaning councillors dominated Council. However many reformists remained in power. In the November 2003 elections, former councillor David Miller won against conservative John Tory in a tight race. Reformists resumed their dominance at City Hall, to the great relief of many city employees. David Miller does not claim a reformist label, but his approach and politics are very much in-tune with the City's political culture. His reformism, however, has explicitly incorporated the discourse on global competitiveness.

It becomes increasingly clear that under Mayor Lastman, a pro-growth elitist machine consolidated in Toronto. They focused on the creation of a dense, green, festive, safe, and attractive urban milieu that will bring highly qualified professionals to the city. Reflecting upon amalgamation, Todd reminds us that Toronto's entrepreneurs have supported the idea since the 1970s, as it simplifies zoning regulations and destabilises reformists (Todd 1998). This business support for amalgamation became evident in 1997, when the Transition Team was appointed by the provincial government to organise amalgamation. Many business leaders were appointed to the Team (Todd 1998). After amalgamation, Toronto's political agenda was dominated by pro-growth development. This agenda was further supported by provincial policies such as the Safe Street Act criminalising squeegee kids and panhandling, the legalisation of the 60-hours work week, the loosening of planning codes, the elimination of public housing programs and of rent control, the implementation of a more flexible policy on soil decontamination in order to facilitate redevelopment, and so on. Although Mayor Miller was supported by many of these global elites in his 2003 electoral campaign, a new wind is blowing on Toronto. Miller's first action was to cancel the construction of a bridge to downtown islands, where a small airport was scheduled for expansion. This is a sign of a different kind of growth, more akin to reformism. However, Miller does not reject growth altogether. He is very clear on his global dreams for the city.

Conclusion

The reformist urban and democratic heritage has largely been taken up by the newly amalgamated city through its insistence on density, quality of life, beautification, racial and cultural harmony, diversity, participation, the environment, and so on. The rise to power of a new urban regime that could be qualified as neo-liberal, has built on this specific political culture and a discourse opposing the urban and the suburban. While in 1998, in the midst of intense mobilisation against amalgamation, it seemed that the conservative provincial government had underestimated the strength of the reformist political culture, it is now more appropriate to speak of a new synthesis through which the new city did not follow the path of low-density suburban development as it was predicted by C4LD. Instead, the new city incorporated this reformist political culture into a neo-liberal regime, in-line with provincial policies, while redefining what urbanity and diversity mean.

While in the 1970s the 'evil' for reformists was suburban sprawl (understood through a simplistic duality between urban diversity and suburban homogeneity), leftist movements today have to go beyond this dichotomy in order to incorporate socio-demographic transformations occurring in Toronto as elsewhere. The 'enemy' is no longer the suburbs (which are themselves diversifying and urbanising), but instead the vision of urbanity promoted by the neo-liberal growth model based on urban density. The importance of cities in the global economy has been largely documented (Castells 1989; Sassen 1994). A dynamic, festive, multicultural, central city offers a clear competitive advantage for city boosters. This vision of urbanity, cleared of the unexpected, of undeveloped lands, of the undesirable and other street behaviour (graffiti, alternative music, etc.) attempts, to use Zukin's formulation, to 'pacify with cappuccino' (Zukin 1995; Smith 1996).

C4LD and reformists have much to gain in abandoning this false dichotomy between the centre city and the suburbs. A critique of the neo-liberal vision of urbanity has to include alliances with suburban residents. The new local democracy mechanisms implemented by the City of Toronto do not reach out to the exurban communities of the GTA as they were excluded from amalgamation. A regional perspective might be a good starting point for resisting the neo-liberal regime.

New regionalism in Toronto is subject to an ambiguous debate. On the one hand, many provincial and local policy-makers adopted, at the beginning of the 1990s, the regionalist principles promoted in U.S. work (for instance, Ledebur and Barnes 1993). This debate culminated with the publication of the Golden Report, which had been commissioned by the government in power before the rise of Mike Harris. The Golden Report resulted from a task force on the future of the GTA (Greater Toronto Area Task Force 1996). When the new Tory government decided to ignore the final recommendations of this task force in 1996, in order to amalgamate only at the scale of Metro Toronto rather than creating a GTA-wide institution, it further sustained this urban-suburban dichotomy. On the other hand, intellectuals and activists were already critical of the Golden Report as it insisted on the necessity for a new GTA regionalism with the goal of increasing competitiveness rather than focusing on social regional development (Todd 1996; Kipfer 1998). Moreover, leftist discomfort with the urban-suburban dichotomy has fuelled an alternative regionalist discourse based on the alleviation of poverty and bioregionalism.

These two forms of regionalism (based on competitiveness or social and ecological principles) remained on the political landscape when, after amalgamation, a movement for more autonomy for Toronto rose out of C4LD, in coalition with some municipal councillors and business philanthropists. It took many forms, from proposing to create a Province of Toronto to modifying the Canadian constitution to create a city-state status for large cities, from pressures to reform the provincial Municipal Act governing the provincial-municipal relations to the elaboration of a GTA Charter that would provide the city-region with a certain level of autonomy from the province (Rowe 2000; Keil and Young 2003). This has now evolved into a national discourse on urban regions, epitomised by the federal government's New Deal for Cities. Again, the rise of urban issues on the federal agenda is led in Toronto by a coalition of business leaders and reformists known as the Toronto City Summit Alliance, another example of a new synthesis between reformism and neo-liberalism (Boudreau and Keil 2004).

To come back to the local democracy mechanisms discussed above, it is not completely fair to attribute the new City's initiatives to a simple response to socio-cultural transformations. While most local actors insist that the City's new participation tools and diversity initiatives are direct responses to studies linking poverty and race (Ornstein 2000), these mechanisms are also built on the will to integrate the reformist political culture into the newly amalgamated city. Hence, the institutional reforms of 1997-98 have in a large part propelled a redefinition of reformism, opening the door to its re-appropriation by a new neo-liberal elite supported by the provincial government. We are thus not faced with a direct response to urban tensions created by socio-demographic changes, but rather with the seizure of an opportunity opened by institutional restructuring. These new local democracy mechanisms exclude 'illegitimate' interlocutors from civil society if they are too threatening for the new regime. Anarchist and alternative globalisation urban social movements, such as squatters, anti-poverty activists, or radical ecologists, are excluded from this approach to local democracy (Barlow and Clarke 2001; Klein 2002).

The case of Toronto clearly illustrates that the discourse on local democracy sustained by reformists has been deradicalised and incorporated into a neo-liberal regime that was built on the opportunities opened by amalgamation and cutbacks. As Keil indicated in 1998, the discourse on local democracy elaborated by reformists opposing amalgamation, was accompanied by a less visible diminution of participation in planning decision (Keil 1998). While thousands of residents mobilised by C4LD were demonstrating on the streets of Toronto against amalgamation, important planning decisions were made, detached from public participation: the extension of Pierson International Airport, the Air Canada Complex for the hockey club Maple Leafs and the basketball club Raptors, the revitalisation of Dundas Square in a similar fashion as New York's Time Square, etc. These planning decisions have a definitive impact on urban space and its access for disadvantaged people. The current discourse on local democracy hides a centralisation of power underway in Toronto since the rise of a neo-liberal regime (Keil 1998).

The discourse on local democracy in Toronto is not unique; rather, it displays a characteristic neo-liberal re-appropriation of urbanity and diversity. As Jessop suggests, neo-liberalism is part of a continuum in which neo-communitarianism (of the Toronto reformist type) can be seen as a form of neo-liberalism in that local democracy mechanisms designed to counter neo-liberalism often lead to policies very similar to that of neo-liberalism (decentralisation, public-private and public-community partnerships, less stringent bureaucratic rules, promotion of the service and social economy, etc.) (Jessop 2002). Neo-communitarianism and the discourse on local democracy in Toronto carry a double meaning, which calls for a profound rethinking of reformism. The November 2003 election of Mayor Miller, combined with the landslide defeat of the provincial conservatives in the October 2003 elections, may be a good opportunity for undertaking a critical rethinking of reformism. While the left senses a wind of freedom blowing over Toronto after a decade of anti-statist neo-liberalism, it remains worried about the social consensus on the necessity to focus on global competitiveness. It remains to be seen whether the most critical elements of Toronto's left movements will take the forefront or whether it will simply reproduce urban middle class reformism and tap itself on the back, proclaiming how wonderful our city is.

References

Allen, M. (1997) *Ideas That Matter: The Worlds of Jane Jacobs*. The Ginger Press Owen Sound.

Barlow, M. and T. Clarke (2001) *Global Showdown: How the new activists are fighting global corporate rule*. Stoddart Toronto.

Boudreau, J.-A. (1999) Megacity Toronto: Struggles over Differing Aspects of Middle-Class Politics, *International Journal of Urban and Regional Research*, 23.4, pp.771-781.

Boudreau, J.-A. (2000) *The Megacity Saga: Democracy and Citizenship in This Global Age*. Black Rose Books, Montreal.

Boudreau, J.-A. (2003) Questioning the Use of 'Local Democracy' as a Discursive Strategy for Political Mobilization in Los Angeles, Montreal and Toronto. *International Journal of Urban and Regional Research*, 27.4, pp.793-810.

Boudreau, J.-A. and R. Keil (2004) *In search of a new political space? City-regional institution-building and social activism in Toronto*. Prepared for the Annual Meeting of the Association of American Geographers, Philadelphia.

Castells, M. (1983) *The City and the Grass-roots: A Cross-cultural Theory of Urban Social Movements*. Edward Arnold, London.

Castells, M. (1989) *The Informational City*. Blackwell, Oxford.

Caulfield, J. (1988a) Canadian urban 'reform' and local conditions: an alternative to Harris's 'reinterpretation', *International Journal of Urban and Regional Research*, 12, pp.477-484.

Caulfield, J. (1988b) 'Reform' as a Chaotic Concept: The Case of Toronto, *Urban History Review/Revue d'histoire urbaine*, XVII.2, pp.107-111.

Caulfield, J. (1994) *City Form and Everyday Life: Toronto's Gentrification and Critical Social Practice*. University of Toronto Press, Toronto.

City of Toronto (1997) *Community Consultation: Options for Decision Making in the Megacity*. City of Toronto, Toronto.

City of Toronto (1999) *Toronto Plan*. City Planning Division, Urban Development Services, City of Toronto, Toronto.

City of Toronto (2000) *Toronto Plan: Directions Report*, City Planning Division, Urban Development Services, City of Toronto, Toronto

City of Toronto (2001) *Development of a City of Toronto Declaration and Plan of Action Regarding the Elimination of Racism in Relation to the United Nations – World Conference Against Racism, Racial Discrimination, Xenophobia and Related Intolerance (UN-WCAR)*, downloaded from the internet on February 20, www.city.toronto.on.ca.

City of Toronto (2002) *Plan of action for the elimination of racism and discrimination*. Downloaded from the internet on February 19, www.city.toronto.on.ca.

City of Toronto (2003) *2003 City Budget Community* Workbook. Downloaded from the internet on 21February 2004, www.city.toronto.on.ca.

City of Toronto (2004) *Listening to Toronto: Presentation to the Joint Meeting of the Policy and Finance Committee and Budget Advisory Committee*. City of Toronto, Toronto.

Community Social Planning Council of Toronto (2000) *An Act for the new Millennium?* Community Social Planning Council of Toronto, Toronto.

Conway, J. (2004) *Identity, Place, Knowledge: Social Movements Contesting Globalization*. Fernwood Publishing, Toronto.

Croucher, S.L. (1997) Constructing the Image of Ethnic Harmony in Toronto, Canada: The Politics of Problem Definition and Nondefinition, *Urban Affairs Review,* 32.3, pp.319-347.

Filion, P. (2000) Balancing concentration and dispersion? Public policy and urban structure in Toronto, *Environment and Planning C: Government and Policy,* 18, pp.163-189.

Greater Toronto Area Task Force (1996) *Greater Toronto: Report of the GTA Task Force.* Publications Ontario, Toronto.

Harris, R. (1987) A social movement in urban politics: a reinterpretation of urban reform in Canada, *International Journal of Urban and Regional Research,* 11, pp.363-379.

Harris, R. (1988) The interpretation of Canadian urban reform: a reply to Caulfield, *International Journal of Urban and Regional Research,* 12, pp.485-489.

Jacobs, J. (1961) *The Death and Life of Great American Cities.* Vintage Books, New York.

Jessop, B (2002) Liberalism, Neo-liberalism and Urban Governance: A State-Theoretical Perspective. In N. Brenner and N. Theodore (eds) *Spaces of Neoliberalism; Urban Restructuring in North America and Western Europe,* Blackwell, Malden MA..

Keil, R. (1998) Toronto in the 1990s: Dissociated Governance?, *Studies in Political Economy,* 56, pp.151-167.

Keil, R. and D. Young (2003) A Charter for the People? A Research Note on the Debate About Municipal Autonomy in Toronto, *Urban Affairs Review,* 39.1, pp.87-102.

Kipfer, S. (1998) Urban Politics in the 1990s: Notes on Toronto. In I. Zurich (ed) *Possible Urban Worlds,* ETH, Zurich.

Kipfer, S. (1999-2000) Whose City Is It? Global politics in the Mega-City, *Cityscope: Newsletter of the Community Social Planning Council of Toronto,* pp.13-17.

Kipfer, S. and R. Keil (2002) Toronto Inc.? Planning the Competitive City in the New Toronto, *Antipode,* 34.2, pp.227-264.

Klein, N. (2002) *Fences and Windows: Dispatches from the front lines of the globalization debate.* Vintage Canada, Toronto.

Ledebur, L.C. and W.R. Barnes (1993) *All in together.* National League of Cities, Washington, DC.

Lemon, J.T. (1996) Toronto, 1975: The Alternative Future. In J.T. Lemon (ed) *Liberal Dreams and Nature's Limits: Great Cities of North America Since 1600.* Oxford University Press, Toronto.

Lorimer, J. (1970) *The Real World of City Politics.* James Lewis and Samuel, Toronto.

Milroy, B.M. (2000) Toronto's Legal Challenge to Amalgamation. In C. Andrew, K. Graham and S. Phillips (eds) *Urban Affairs: Is it Back on the Policy Agenda?/Les affaires urbaines: sont-elles de nouveau à l'ordre du jour?.* McGill-Queen's University Press, Montreal and Kingston.

Ornstein, M. (2000) *Ethno-Racial Inequality in the City of Toronto: An Analysis of the 1996 Census.* Access and Equity Unit, Strategic and Corporate Policy Division, Chief Administrator's Office, Toronto.

Rowe, M.W. (2000) *Toronto considering self-government.* The Ginger Press, Inc. Owen Sound, Ontario.

Sassen, S. (1994) *Cities in a World Economy.* Pine Forge Press, Thousand Oaks.

Sewell, J. (1972) *Up Against City Hall.* James Lewis and Samuel, Toronto.

Sewell, J. (1993) *The Shape of the City: Toronto Struggles with Modern Planning.* University of Toronto Press, Toronto.

Sewell, J. (2000) The City Status of Toronto. In M.W. Rowe (ed) *Toronto Considering Self-Government.* The Ginger Press, Inc., Owen Sound.

Smith, N. (1996) *New Urban Frontier: Gentrification and the Revanchist City.* Routledge, New York.

Task Force on Community Access and Equity (1998) *Consultation Guide.* City of Toronto, Toronto.

Task Force on Community Access and Equity (2000) *Diversity Our Strength, Access and Equity Our Goal.* Downloaded from the internet on www.city.toronto.on.ca

Todd, G. (1996) *Under the big top: internationalization and the regional city state.* Annual Meeting of the Canadian Political Association, Brock University, St.Catharine's.

Todd, G. (1998) Megacity: Globalization and Governance in Toronto, *Studies in Political Economy,* 56, pp.193-216.

Toronto City Council Policy and Finance Committee (2001) *Diversity Advocate.* City of Toronto, Toronto.

Zukin, S. (1995) *The Cultures of Cities.* Blackwell, Cambridge.

Chapter 8

Municipal Reform and Public Participation in Montreal's Urban Affairs: Break of Continuity?

Anne Latendresse

Introduction

For a number of years, there has been renewed interest in local democracy and, by extension, in issues such as inclusiveness, citizenship and participation in the city. Not so long ago, the large city, or the metropolis, was still associated with, in Simmel's words, a place of anonymity, even disorder, chaos, and social problems. However, in the more recent literature, the city, in a context of economic and cultural globalisation, has been increasingly viewed as a privileged space where major innovations in democracy and public affairs management take place. In fact, as the state undergoes changes, the city is fostering the emergence of new modes of relations between the various types of actors which involve negotiation, partnership and mediation (Le Galès 1995). Still others consider that the large city constitutes a space for the renewal of democracy and citizenship. Thus, in referring to Berlin's experience, J. Friedman maintains that the metropolis can become the place for the expression of a local citizenship (Friedman 1998).

However, not everybody agrees with these views. For Perrineau (2003), modern societies are confronted with the paradox of democratic disillusionment. On the one hand, representative democracy is being questioned and even challenged (Barber 1997; Courtemanche 2003). For Norris (1999), a 'growing cynicism' towards democratic governments explains the decline in democratic participation (electoral participation, partisan activism and civic engagement). In Montreal, for example, the participation rate in the most recent municipal elections was only 48 per cent, lower than the rate in previous elections. Yet the 2001 elections involved major democratic issues for Montrealers. On the other hand, a quest for new spaces of participation in the management of public affairs is emerging. This demand for a strengthening of democracy and greater citizen participation in the definition of the city is expressed, in Quebec, within a historical context that dates back to the 1960s (Hamel 1991).

Just like Quebec's other main cities, Montreal has recently undergone an unprecedented municipal re-organisation. As the result of the merging of the former central city and 27 suburban municipalities of the island, the birth of the new City of Montreal has given Montrealers the opportunity to consider the future of their city.

Following the initiative of a community organisation, the *Société de développement communautaire de Montréal* (SODECM), two Citizens' Summits on the Future of Montreal were held in 2001 and 2002 respectively, bringing together hundreds of people from numerous sectors and milieus (union, community, feminist, ecological, cultural communities of the Island of Montreal). This was a first for Montreal. For the organisers of these initiatives, the municipal reform, despite its shortcomings and the fact that it was imposed by the provincial government, represented an opportunity to strengthen local democracy and promote the creation of new participation spaces, in particular at the borough level. Indeed, Montrealers and, in particular, community organisation activists engaged in urban issues had been complaining for a number of years that the municipality was suffering from a serious democratic deficit.

This chapter will examine local democracy in the Montreal context since the municipal reform. We will address the following questions: To what extent is the creation of the new city an opportunity for democratic strengthening of the city's affairs? Does the establishment of new structures and institutions mark a break with the previous situation or is it in continuity with the previous situation? More specifically, to what extent can the public consultation mechanisms and bodies respond to society's demand for greater participation? Lastly, how will Montrealers respond to this institutional reconfiguration?

A Managerial Approach to Municipal Reform

The Government of Quebec has imposed an unprecedented municipal re-organisation whose stated goal was to reconfigure the Quebec municipal scene. On several occasions since the 1960s, the provincial government has mandated different commissions to examine the issue of municipal re-organisation (Collin 1999; Desrochers 2003; Hamel 2001). As Desrochers has pointed out, Montreal was not an exception and 'the majority of the reform projects which were developed were not followed up – the best known being unquestionably the project embodied by the *Commission de développement de la Métropole*, an organisation which was stillborn in 1997' (Desrochers 2003, p.30) (translation).

In the eyes of the Quebec government, led by the *Parti Québécois* (PQ) from 1994 to 2003, the metropolitan agglomeration had issues that were both similar to those of other large cities of Quebec and also specific to itself. Indeed, the Montreal agglomeration is unique in several respects. Approximately half of the total population of Quebec and 1.7 million jobs are concentrated there. Its annual production of $86 billion represents nearly half of Quebec's jobs and GDP.[1] In intercultural terms, 45 per cent of immigrants in Quebec are concentrated in the metropolitan region (Charbonneau and Germain 2003, p.316), a proportion that rises to 70 per cent for the Island of Montreal as a whole (Germain 2003, p.79). The use of the French language on the island is declining. In the historical context of Quebec where the majority of the francophone population lives in a broader political and cultural environment that is mainly anglophone, the ethno-linguistic composition of

---

[1] Government of Quebec. La réorganisation municipale. Changer les façons de faire, pour mieux servir les citoyens, p.37.

the population living in the territory and the political-administrative division thus have an important political dimension (Germain 2003; Serré 2002). This question constituted an important issue since, as recalled by J.-R. Sansfaçon, editor-in-chief of the daily newspaper *Le Devoir*, Montreal anglophones 'viewed their municipalities as their primary bastions that ensure the protection of their linguistic interests and protect them against the ambitions of a higher-level government which would not respect their vision of the country' (Sansfaçon cited in Serré 2000) (translation).

In its White Paper on Municipal Reorganisation (*La réorganisation municipale : Changer les façons de faire pour mieux servir les citoyens*), the provincial government identified several problems which, in its view, hampered the development of Montreal: urban sprawl, municipal fragmentation, fiscal inequity and the discrepancy between the territory covered by the *Communauté urbaine de Montréal* (CUM, Montreal Urban Community, the metropolitan management and planning authority that existed from 1970 until the reform) and the actual territory in which the population lives and activities take place, which covers a much larger area. The CUM covered an area of 502 km², or only 13 per cent of the territory of the census metropolitan region. The CUM was inhabited by a total population of 1.8 million or 53.5 per cent of the population of the Census Metropolitan Region, estimated to be 3.4 million.

The goal of the PQ government reform was to strengthen the large urban agglomerations and reduce the high number of municipalities in Quebec.[2] Based on this logic, the government opted for the territorial and administrative consolidation of cities in the same agglomeration. Large cities like Montreal, Québec City, and Sherbrooke were merged, leading to the creation of new structures and institutions at the levels of the metropolitan agglomeration, the city and boroughs. Under this municipal re-organisation, the government's main goals emphasised the economies of scale, greater rationalisation in urban management and planning, and the pursuit of better performance in terms of service management.

Like any political project, the reform proposed by the PQ government in the White Paper was based on a number of values. As pointed out by L. Quesnel, 'the municipal re-organisation is not just a matter of administrative re-engineering without repercussions on public action as a whole. On the contrary, the configuration of municipal institutions and their responsibilities, leadership and democratic regime attest to the general policies and normative foundations on which public action is based' (Quesnel 2003, p.3)[3] (translation). Although the White Paper specifies that 'municipal democracy does not amount to a question of bills without affecting the decision-making structures' (p.71) (translation), the main goals stated in fact reflect a managerial approach to municipal territorial administration that strives to make Quebec's cities more efficient and competitive in the context of economic globalisation, thus relegating democratic concerns to the back burner.

---

[2] Quebec had 1306 municipalities in 2000 for a total population of 7,098,298 million.
[3] http://www.vrm.ca/documents/Synthese_demo_parti.pdf consulted on 17 October 2003.

The Suburban Rebellion

In Toronto, opposition to the merger project came from citizens of both the former Toronto central city and the suburban municipalities (see chapter by Boudreau). In contrast, in Montreal, the discontent mainly came from the suburban municipalities which were part of the Montreal Urban Community (CUM). During the consultation process launched by the government, the mayors of the suburban municipalities soon let it be known that they were opposed to the merger option (Sancton 2000). Fifteen of the 27 mayors of the suburban municipalities had refused the invitation to participate in the Mayors' Council set up by the Montreal Transition Committee which was mandated to implement the new City of Montreal as of 1 January 2002.

Nineteen of the 27 suburban municipalities contested the legitimacy of the government's action in the courts. At the same time, a coalition of citizens and citizen committees called *Démocracité* was set up to oppose the merger. This coalition carried out vast public mobilisation campaigns to denounce the PQ's project. These opponents were concerned about a tax increase and deterioration of services. Their demands also related to the preservation of their identity (in particular, linguistic). Although the PQ government's proposal stipulated that the former municipalities could preserve their bilingual status, opponents from the overwhelmingly English-speaking suburbs feared that this provision was not enough. In fact, the Charter of Montreal stipulates that Montreal is a French-speaking city. Third, several opponents to the merger argued that the small size of their municipality and their closeness to their elected officials guaranteed them greater democracy. The creation of a large city made them worry about the loss of identity, bureaucratisation and poor management of the city's affairs.

Except for the mobilisation campaigns in the suburbs, the merger project did not arouse much opposition from the rest of the population. The community movement, for example, the *Front d'action en réaménagement populaire* (FRAPRU) or the *Table des regroupements des organismes en éducation populaire*) did not oppose the merger but emphasised that the latter should foster equity in terms of services, equipment and cost sharing between the wealthier and the less wealthy areas of the island. However, in general, the residents had little to say in the public debate which was largely dominated by the 'anti-merger' faction.

The bill proposed by the PQ was finally adopted on 20 December 2000. At the municipal elections in the merged cities in 2001, new political parties were created, amalgamating various political forces, including the anti-merger movement. In Montreal in particular, several former suburban mayors opposed to the mergers were elected to municipal council as members of Gérald Tremblay's *Union des citoyens et des citoyennes de l'île de Montréal* (UCIM, Montreal Island Citizens Union Party), which won the municipal elections of 4 November 2001.

On 14 March 2003, the PQ was ousted by the Quebec Liberal Party (PLQ) which made its opposition to the mergers part of its electoral campaign platform. The PLQ even proposed implicitly to demerge the cities if it was elected. A few months after its election, the new government tabled a new bill (Bill 9), asking the new cities to submit their re-organisation plans. This measure opened the door to the 'demergers' and the debate has been going on ever since.

One Island, One City: Democracy with Variable Geometry

The issues of democracy related to the reform cannot be discussed without addressing the aspect of representation and powers. The merger option retained for Montreal was based on the administrative and territorial consolidation of 28 municipalities of the Island of Montreal. Two different political-administrative divisions have since defined the territorial entities of the new city. While bringing together the former municipalities, the Government of Quebec respected the pre-existing boundaries as the principal criterion used to mark out the territory. The nine boroughs of the former City of Montreal were maintained and 18 new boroughs were created out of the 27 suburban municipalities.

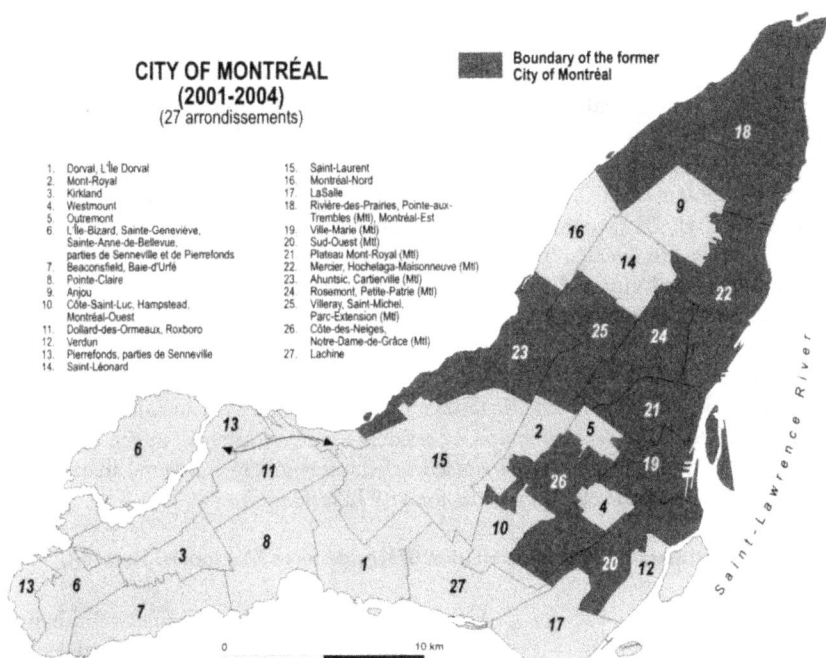

CITY OF MONTRÉAL
(2001-2004)
(27 arrondissements)

Boundary of the former
City of Montréal

1. Dorval, L'île Dorval
2. Mont-Royal
3. Kirkland
4. Westmount
5. Outremont
6. L'île-Bizard, Sainte-Geneviève, parties de Senneville et de Pierrefonds
7. Beaconsfield, Baie-d'Urfé
8. Pointe-Claire
9. Anjou
10. Côte-Saint-Luc, Hampstead, Montréal-Ouest
11. Dollard-des-Ormeaux, Roxboro
12. Verdun
13. Pierrefonds, parties de Senneville
14. Saint-Léonard
15. Saint-Laurent
16. Montréal-Nord
17. LaSalle
18. Rivière-des-Prairies, Pointe-aux-Trembles (Mtl), Montréal-Est
19. Ville-Marie (Mtl)
20. Sud-Ouest (Mtl)
21. Plateau Mont-Royal (Mtl)
22. Mercier, Hochelaga-Maisonneuve (Mtl)
23. Ahuntsic, Cartierville (Mtl)
24. Rosemont, Petite-Patrie (Mtl)
25. Villeray, Saint-Michel, Parc-Extension (Mtl)
26. Côte-des-Neiges, Notre-Dame-de-Grâce (Mtl)
27. Lachine

Figure 8.1   Merger of 28 Municipalities on the Island of Montreal

The boroughs vary greatly in terms of population and territory. According to 2001 census data, Côte-des-Neiges/Notre-Dame-de-Grâce Borough – a borough of the former City of Montreal – is the most populated borough, with 164,350 residents, whereas Dorval-L'île-Dorval, composed of two former municipalities, represents the smallest borough, with 17,332 residents. The average number of people in the boroughs of the former central city is 115,670 compared with only 42,567 in the former suburban municipalities.

This difference would not matter so much if it were not accompanied by an overrepresentation of residents from the suburbs in the decision-making authorities. At the political-administrative level, the reform distinguishes between two types of councillors: city councillors (also called municipal councillors) and borough councillors. The former sit on both the municipal council and the borough council whereas the latter sit on the borough council only. The municipal councillors are elected by the borough electors. They have more power since, despite the decentralisation and sharing of powers, areas of jurisdiction and responsibilities between the central city and the boroughs, the municipal council preserves a number of prerogatives and powers, and makes decisions for the city as a whole. For example, the municipal budget for the city, the loans and the borough budgets in particular are adopted by this authority which is also responsible for revising urban planning.

An examination of the ratio of elected official to elector reveals that residents of the former suburban municipalities are better represented than those of the boroughs of the former central city. The suburb therefore has considerable political weight in the municipal council, with 40 elected officials out of a total of 73 councillors (excluding the mayor) compared with 33 councillors from the former central city. The disproportion is even more evident when we consider that these 40 councillors represent a population of 766,211 versus 33 councillors of the former city for a population of 1,041,031 (according to 1996 statistical data).

Table 1 shows an average of 17,903 electors per city councillor for the boroughs of the former central city and 17,154 electors per elected official for the boroughs created from the former suburban municipalities. The difference is certainly not huge. However, the comparison of the number of electors per elected official does not accurately reflect the representation made by the elected officials. Given the number of residents per elected official, this time including the city councillors and the borough councillors, the representation of residents is twice as small in the boroughs of the former City of Montreal (one elected official for 26,026 residents) than in the former municipalities (one elected official for 11,972 residents).

Table 8.1   Comparison of Representation of Residents to Electors by Borough

	Former City of Montreal	Former suburban e municipalities
Total number of residents	1,041,031	766,211
Total number of electors	766,116	566,095
Average number of residents per city councillor	26,026	23,219
Average number of residents per city and borough councillor	26,026	11,972
Average number of electors per city councillor	17,903	17,154
Average number of electors per city and borough councillor	17,903	8,845

Source: http://www2.ville.montreal.qc.ca/ consulted in August 2003.

In the executive committee, the imbalance in the representation is even more striking. Out of a total of 11 councillors excluding the mayor, seven councillors come from the suburbs whereas only four come from the former central city. Moreover, the chair of the executive committee is now a municipal councillor from Saint-Léonard Borough, a former suburban municipality. When these data are broken down by number of electors per elected official, there are 179,029 electors per elected official for the former City of Montreal versus 80,871 electors per elected official from the former suburban municipalities. Once again, the population of the former City of Montreal is underrepresented.

This overrepresentation of the suburbs within the municipal council results from the fact that the previous government decided to respect the former territorial boundaries, as was done in previous mergers, and thus to preserve the sense of identity of the communities concerned. However, Bill 170 set the number of city councillors for each borough based on the size of the population. Dissatisfied with the number of councillors attributed to them by the bill, elected officials of the former municipalities managed to convince the government to increase this number.

Furthermore, the *Charter of Ville de Montréal* (Annex I of Bill 170 adopted in December 2000) provided for the division of a number of boroughs into electoral districts for the purposes of the election of city councillors or borough councillors. Moreover, it stipulated that the borough councils should be governed by a college of elected officials composed of three to six persons. Fifteen of the 27 boroughs have three councillors only. As a result of numerous mobilisation campaigns by municipalities, groups and citizens, amendments were made in spring 2001 through Bill 29. The number of councillors in the boroughs of Saint-Laurent, Saint-Léonard, Verdun, Montréal-Nord and LaSalle increased from three to five. This exceptional provision, adopted for the first elections only, intensified the disproportion of the political representation in favour of suburban municipalities. For example, there are five councillors (two city councillors and three borough councillors) for a borough like Verdun, which has 60,598 residents, and six councillors for the borough of Côte-des-Neiges/Notre-Dame de Grâce, which has 164,350 residents.

In the case of the executive committee, the political weight given to the suburbs comes from a decision of the Mayor who, under the Charter, has the power to choose the members of this body. To understand the Mayor's choice, the composition of his political party must be taken into account. Created in the context of the merger's implementation, the *Union des citoyens et des citoyennes de l'île de Montréal* (UCIM) stems from an alliance between, on the one hand, Gérald Tremblay, former minister of the Liberal Government in the early 1990s, and on the other hand, the *Rassemblement des citoyens et des citoyennes de Montréal* (RCM) and the elected officials of the suburban municipalities. Pierre Bourque's tenure as mayor of Montreal ended in 1994. As mayor, he had relied on a more striking political leadership approach to strive to keep the former City of Montreal at the centre of the political stage. In contrast, Gérald Tremblay was soon viewed as 'the darling candidate of the suburbs' (*Le Devoir*, 5 November 2001) (translation) since he had succeeded in rallying several mayors in the Montreal region who were opposed to the merger, for example, Peter B. Yeomans of Dorval, Georges Bossé of Verdun, Luis Miranda of Anjou, Bill McMurchie of Pointe-Claire and Frank Zampino of Saint-Léonard. The electoral victory of the UCIM (41 elected councillors out of a total of 73) was essentially a victory for the suburbs.

Although the analysis of political representation clearly shows an imbalance in the representation of interests between the population in the former suburban municipalities and that of the former City of Montreal, what is the division of power within these new authorities?

New Model, New Structures?

At first sight, the institutional model that the Government of Quebec chose for Montreal seemed to be innovative in that it provides for the sharing of powers and responsibilities within a three-tier structure: the *Communauté métropolitaine de Montréal* (CMM, Montreal Metropolitan Community) for the greater metropolitan region, the City Council and the Executive Committee for the city as a whole (i.e. the Island of Montreal) and lastly, the borough councils which are in fact the main innovation of this reform (Bruneault and Collin 1999; Latendresse 2001; Desrochers 2003).

The Metropolitan Agglomeration Level

At the regional level, chapter 34 of Bill 170 provided for the creation of the CMM which started operations on 1 January 2001. This inter-municipal management and planning authority includes 63 municipalities spread over the north and south shores as well as on the Island of Montreal. The territory in which it is authorised to act is considerably more extensive than that covered previously by the CUM. While the CUM territory corresponded to the Island of Montreal, that of the CMM corresponds to the boundaries of the census metropolitan region with an area of 3818 km² and a total population of 3.4 million.

However, this institution cannot be viewed as a supra-municipal government as recommended by the *Groupe de travail sur la région de Montréal* (GTRM, Montreal region working group) (Hamel 2001; Desrochers 2003). This would have implied that it has taxation powers and its political staff are directly elected by the population. Nevertheless, the CMM incorporates new responsibilities and new powers. It must develop a metropolitan plan for land use and urban planning. It must also 'establish a strategic vision of economic, social and environmental development for the metropolitan region and ensure that the urban development plan corresponds to this vision'[4] (translation). In addition to jurisdiction in the four areas of land-use planning, economic development, waste management planning and public transportation (shared with the municipal council and the borough councils), the CMM also supervises artistic and cultural development, equipment and infrastructures management, social housing and international promotion (Desrochers 2003).

The CMM is made up of 28 representatives from the Montreal metropolitan region which includes the Island of Montreal, Laval, Longueuil, the North Shore and the South Shore. Its membership includes the mayor and 13 city councillors from the City of Montreal, the mayor and two municipal councillors from the City of Laval, the mayor and two municipal councillors from the City of Longueuil, four mayors from

---

[4] CMM Web site http://www.cmm.qc.ca consulted on August 17, 2003.

the North Shore, and four mayors from the South Shore. This authority also has a committee made up of eight members from the CMM Council who are divided as follows: the mayor and three city councillors from the City of Montreal, the mayors of Laval and Longueuil and two other mayors, one from the South Shore and one from the North Shore. The City of Montreal chairs the CMM and has the largest representation within the CMM Council and its commissions because it groups together 53 per cent of the total population of the metropolitan region.

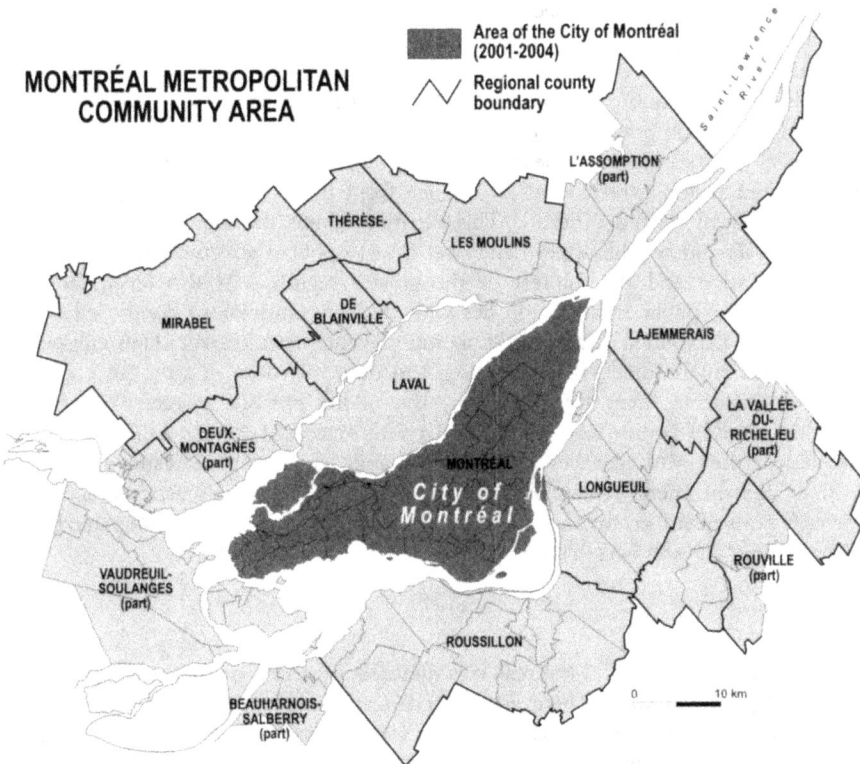

Figure 8.2 *La Communauté Métropolitaine de Montréal* (Montreal Metropolitan Community)

To assess the significance of the CMM's implementation, it is important to note that it succeeds a number of metropolitan bodies such as the *Commission métropolitaine de Montréal* (Metropolitan Commission of Montreal) established in 1921, the *Corporation du Montréal métropolitain* (Corporation of Metropolitan Montreal) created in 1959, and the CUM which was set up in 1970. According to M.-O. Trépanier, 'it was difficult for these different bodies to survive the squabbles between Montreal and the suburbs. The smooth functioning of the MUC itself will always suffer from the disproportion between Montreal and the CUM suburbs.' (Trépanier 1998, p.326) (translation). Moreover, the implementation of the CMM does not seem to have solved the issue of municipal fragmentation and the great number of authorities which take action in this territory. It should be pointed out that the regional county municipalities (RCMs) established on the island in 1980 are still in place as well as institutions such as the local and regional school boards, the health and social services boards, the regional development councils, transportation corporations, etc. The presence of numerous institutions taking action in this territory makes it difficult to foster collaboration and activates the mechanisms of isolation, defence or competition (Trépanier 1998).

The implementation of the CMM is in continuity with the previous situation. In fact, the Government of Quebec could have taken advantage of this event to create a true metropolitan government authority and give it concrete means to ensure better planning at the level of the greater metropolitan region. However, it opted to maintain the status quo. The CMM is still an inter-municipal co-ordination authority deprived of taxation power. Moreover, on the democratic level, the CMM is made up of councillors from municipalities who are not directly elected by the residents of the metropolitan region, which keeps a distance from the population and weakens the accountability of elected officials. This situation allows the municipalities to assert their interests and get the upper hand over a metropolitan government that is based on co-operation and the interest of the greater region. At this level, the only mechanism for public consultation lies with the six commissions composed of the members of its council (planning, economic development, metropolitan equipment and finance, environment, social housing, and transportation). Lastly, since citizens have no influence over this authority, the CMM remains vague, abstract and misunderstood by people who are unfamiliar with municipal affairs. As F. Desrochers concluded: 'the CMM, like the CUM before it, appears to be more destined to remain a 'thing' of the member-municipalities, a useful forum for discussions, accessible through convenient co-optation; a select club for mayors only, accompanied by a planning role' (Desrochers 2003, p.34) (translation).

The City Level

For its part, the City of Montreal is structured by three political authorities: the municipal (or city) council, the executive committee and lastly, the borough councils. Except for the creation of borough councils which constitute the main novelty of the reform, the city council and the executive committee are in continuity with previous structures. The city council, which is composed of the mayor and 73 city councillors, has jurisdiction in the following areas: land use and urban planning, community and social development, municipal road system, production and distribution of drinking water, water purification, recovery and recycling of residual materials, social housing, etc.[5] Unlike the city council of the former City of Montreal in which all the powers and services were centralised, the new city council shares some of its powers and responsibilities with the borough councils which have become political-administrative entities responsible for all affairs relating to their respective boroughs. However, the Charter specifies clearly that the preparation of budgets, the adoption of the urban

---

[5] Ville de Montréal Web site http://www2.ville.montreal.qc.ca consulted on June 10, 2003.

planning program as well as human resources management in particular (including employees of the borough offices) remain the prerogatives of the city council which may oppose the decisions made by the borough councils. In the current context where the Liberal Party government will allow for new municipal re-organisations which can lead to a demerger, the city administration adopted a decentralisation plan in September 2003, which granted more powers to the boroughs. Mayor Gérald Tremblay thus hoped to ensure the future of the large city by meeting the expectations of the councillors of the former suburban municipalities who demanded greater powers for their boroughs.

For its part, the executive committee is still an authority which has considerable power. It prepares and submits to the city council the city's annual budget, including the loan projects. In addition, the council is authorised to sign contracts under $100,000. Although the executive committee's meetings are closed to the public, a few meetings were held in public in fall 2002 at the request of a city councillor.

As regards citywide public consultation, the city council gathers the opinions of the public during the council's meetings. Moreover, the law provides for the creation of a public consultation office whose mandate is 'to propose a regulatory framework for the public consultations carried out by the official of the city in charge of such consultations so as to ensure the establishment of credible, transparent and effective consultation mechanisms.'[6] Also regarding consultation, the city council has set up ten permanent commissions, each responsible for one of the following areas: the council of the chair, finance, urban planning, land use planning, environment and sustainable development, economic development, intercultural relations, housing and social and community development. In addition, the City has an ombudsman.

It should be recalled that similar public consultation bodies had been introduced in the years when the municipality was led by the RCM, between 1988 and 1994. This party's arrival in power, through the support in particular of the labour movement and urban social movements, had marked a major turning point in the city's administration which, for the previous 30 years, had been led by an authoritarian and populist mayor. This RCM victory in 1986 and the adoption in 1988 of a public consultation framework policy constituted an important step for the modernisation of the political-administrative apparatus (Hamel 1991). At the time, the administration set up three types of bodies under the policy: the *Bureau de consultation publique de Montréal* (BCM, Montreal Public Consultation Bureau), the council's five permanent commissions and the district council-committees. Except for a very few differences, the council's permanent commissions and the *Office de consultation de Montréal* established since the reform are very similar to them.

Scope and Limitation of the Borough Councils

The creation of the borough councils undoubtedly constitutes the main innovation of this reform. For the first time in the history of Montreal, its elected officials (city councillors and borough councillors) can exert certain powers and make decisions relating to the management, planning and affairs related to the community and local

---

6 http://www2.ville.montreal.qc.ca/vie_democratique/consultation_publique.htm consulted on June 10, 2003.

development in their borough's territory. The borough councils are responsible for the management of local services, equipment and infrastructures as well as the social, community and local development, parks and the road system of the borough. They also have a budget determined by the central administration based on an equalisation formula which must, in principle, help to restore the disparities between boroughs. Under the Charter, the boroughs cannot levy taxes (which is being demanded by a number of elected officials in the current debate on decentralisation). However, the law allows them to levy a tax to provide a special service. The decentralisation plan adopted recently by the administration will effectively increase the boroughs' power, in particular in tariffing.

The borough has a borough director and employees responsible for the main services. The borough councils have set up means of communication and information for residents of their respective boroughs and hold monthly meetings in which the public can express their opinions. Each borough is administered by a council composed of a chair and city or borough councillors, depending on the size of the population. The councillors are elected by the electoral districts.

Although the law provides for public consultation bodies and mechanisms on issues related to urban and city planning, this is not provided at the borough level, except for urban planning council committees (*comités-conseils d'urbanisme*, CCUs). These council committees are composed of at least one elected official from the borough and experts or citizens whose duty is to examine the projects submitted and make recommendations to the elected officials of the borough council. Their composition and their functioning vary from borough to borough. In the majority of boroughs, their meetings are held in camera. However, in some boroughs, for example in Saint-Laurent, the CCU sits in public and citizens can participate.

Pursuant to the *Cities and Towns Act*, the borough councils can mandate any borough authority to conduct consultations on areas over which the borough has jurisdiction. They are also authorised to form standing commissions, which has led to the establishment of working commissions or committees in many boroughs. Once again, there is a diversity of cases. The commissions work in areas as varied as leisure, culture, public works, transportation, economic development, public security, the environment and sustainable development. Each borough established its own functioning and standards. In the majority of boroughs, members of the commissions are designated by the elected officials. However, in the cases of Mont-Royal and Westmount, members are recruited through a public call for candidates.

The creation of borough councils seems to have been well received by the citizens of the former City of Montreal who, for the first time in the history of their city, enjoy local management. However, under the *Cities and Towns Act*, the former suburban municipalities were given full powers associated with the municipalities of Quebec (other than those of Montreal and Québec City which had their own charter): taxation, adoption of budget, resource allocation, adoption and revision of their urban planning program, etc. However, the integration of these suburban municipalities into the new city led to a change in their legal status. From municipalities with full powers, they were transformed into boroughs with more limited powers. The imposition of standardised structures on the boroughs of the entire new city has thus had different impacts, depending on the previous legal status. Once again, the term 'democracy with variable geometry' can be used to characterise the situation since we are faced with an undeniable territorial differentiation in the organisation of democratic life. In fact, the elected officials of the boroughs of the former suburban municipalities had to deal with a loss of autonomy and areas of jurisdiction which was nevertheless compensated by a large political representation within the city council and the executive committee. In the case of the boroughs of the former central city, this reform constitutes an undeniable gain in terms of powers but a weakening of their representation within the institutional structures represented by the city council and the executive committee.

The new institutional arrangement resulting from the harmonisation of different institutional structures set up at the levels of the metropolis, the city and boroughs is thus in continuity with the previous situation. Although the CMM's territory has been enlarged and new responsibilities have been added, it is still directly connected with the former CUM. At the city level, the *Office de consultation de Montréal* strongly resembles the *Bureau de consultation publique* of the RCM era. The only innovative element thus lies in the creation of the borough councils which have a budget and powers linked to the management of local services. It remains to be seen to what extent the elected officials and the citizens will participate in these new spaces.

Conclusion

The municipal reform in Montreal has thus had different impacts depending on the various sectors of the island. Based on the neo-institutionalist approach, which takes into consideration past institutional legacy (Hall and Taylor 1997), we can hypothesise that the territorial differentiation in terms of organisation of local democracy is maintained. In fact, it is legitimate to presume that the democratic practices of both elected officials and citizens are still imbued with the legacy of previous municipal structures.

This hypothesis is supported by the current debate on the decentralisation and the demerging advocated by a great number of municipal councillors who represent the boroughs of the former suburban municipalities. These councillors, many of whom were mayors of their municipality, are expressing a desire to recover the powers they previously had. They are determined to obtain more powers and jurisdiction areas from the central city. Although it is still too early to confirm these results, a current survey indicates that the borough councils are attempting to maintain the consultation bodies and the working committees inherited from the time of the former municipalities. In other words, their goal of returning to the pre-reform situation has led them to opt for a strategy aimed at preserving previous institutional structures and practices.

The situation of the boroughs of the former central city is different. The institutional context has changed and this marks a break with the previous situation since councillors can now administer the boroughs, and is in continuity with the previous situation since the consultation bodies are highly similar to those that had been set up by the RCM. For the councillors of the former central city, and in particular those of the UCIM, the reform has created a space that even allows some borough councils to innovate and perhaps thus to meet citizens' expectations regarding participation. For example, in the first year of its existence, the council of

Côte-Saint-Luc, Notre-Dame-de-Grâce Borough presented the budget to the public. This year, it organised two special meetings on the budget to allow organisations and citizens to voice their opinions on the allocation of budgetary resources. For its part, the council of Plateau Mont-Royal has mandated a working committee to examine and make concrete proposals on participative democracy. Decentralisation has thus opened up a space where it is possible to innovate at the borough level. At the same time and paradoxically, it encourages the former suburban municipalities to maintain the status quo. This democracy with variable geometry can lead to a situation whereby, in the same city, some boroughs would be open and participative while others would remain traditional bastions held by local figures.

References

Barber, B. (1997) *Démocratie forte*. Paris, Desclée de Brouwer.

Boudreau, J.-A. (2000) *The MegaCity. Democracy and Citizenship in this Global Age*. Montréal, Black Rose Books.

Bruneault, F. *et al.* (2001) *Les structures de représentation politique*. Document published as part of the project entitled 'Démocratie municipale à Montréal: Des clefs pour analyser les enjeux de la réforme'. Web site, Villes, régions, monde. Villes, régions, monde.http://www.vrm.ca/démocratie_capsule

Bruneault, F. and J.-P. Collin. (2001) *Le partage des compétences*. Document published as part of the project entitled 'Démocratie municipale à Montréal: Des clefs pour analyser les enjeux de la réforme'. Web site, Villes, régions, monde.http://www.vrm.ca/démocratie_capsule

Charbonneau, J. and A. Germain (2002) Les banlieues de l'immigration, *Recherches sociographiques*, Vol. XLIII, No.2, pp.311-328.

Collin, J.-P. (1998) La création de la CUM en 1969: circonstances et antecedents. In Y. Bélanger *et al.* (ed) *La CUM et la région métropolitaine. L'avenir d'une communauté*. Sainte-Foy, Presses de l'Université du Québec, pp.5-17.

Collin, J.-P. (1999) Quel modèle de gestion métropolitaine pour les villes-régions canadiennes? In C. Andrew *et al.* (eds) *Les villes mondiales. Y a-t-il une place pour le Canada?* Ottawa, Presses de l'Université d'Ottawa, pp.403-420.

Courtemanche, G. (2003) La Seconde Révolution tranquille, Montréal, Boréal.

Desrochers, F.G. (2003) La réforme est un long fleuve tranquille, *Possibles*, Vol.27, No.1-2, pp.28-36.

Friedman, J. (1998) The New Political Economy of Planning: The Rise of Civil Society. In M. Douglass *et al.* (eds) *Cities for Citizens. Planning and The Rise of Civil Society in a Global Age*. Chichester; John Wiley and Sons.

Germain, A. (2003) Deux Montréal dans un? *Possibles*, Vol.27, No.1-2, pp.78-88.

Hall, P.A. and Taylor, R.C.R. (1997) La science politique et les trois néo-institutionnalismes, *Revue française de science politique*, Vol.47, No.3-4, pp.469-496.

Hamel, P. (1991) *Action collective et démocratie locale. Les mouvements urbains montréalais*. Montréal, Les Presses de l'Université de Montréal.

Hamel, P. (1997) Démocratie locale et gouvernementalité: portée et limites des innovations institutionnelles en matière de débat public. In M. Gariépy, M. and M. Marié (eds) *Ces réseaux qui nous gouvernent?* Paris, L'Harmattan, pp.403-424.

Hamel, P. (2001) Enjeux métropolitains: les nouveaux défis, *International Journal of Canadian Studies*, No.24, pp.105-127.

Ministère des Affaires municipales et de la Métropole, (2000) *La réorganisation municipale: changer les façons de faire pour mieux servir les citoyens*. Québec, Government of Quebec.

Norris, P. (ed) (1999) *Critical Citizens. Global Support for Democratic Governance.* Oxford, Oxford University Press.

Perrineau, P. (2003) *Le désenchantement démocratique.* La Tour d'Aigues, L'Aube.

Quesnel, L. (2003) Territorialisation de l'action publique. Efficacité et équité. Quelques éléments de réflexion. Available on the Web site Villes, régions, monde.

Sancton, A. (2000) *La frénésie des fussions. Une attaque à la démocratie locale.* Montréal, McGill-Queen's University Press.

Serré, P. (2000) Fusion des municipalités sur l'île de Montréal. Une réforme technocratique sans dimension démocratique? Web site http://www.vigile.net/00-12/fusions-serre.html

Trépanier, M.-O. (1998) Les défis de l'aménagement et de la gestion d'une grande région métropolitaine. In Manzagol C. and Bryant C. *Montréal, 2001. Visages et défis d'une métropole.* Montreal, Les Presses de l'Université de Montréal, pp.319-340.

Chapter 9

The Fight for Jobs and Economic Governance: The Montreal Model[1]

Jean-Marc Fontan, Juan-Luis Klein and Benoît Lévesque

Introduction

The aim of this chapter is to analyse the importance of voluntary initiatives taken by the social actors in what we refer to as abandoned or marginalised districts. The goal of these initiatives is to mobilise varied resources in order to soften what are seen by the communities living in these districts, as the 'destructuring' effects of globalisation. These actions seek, in some cases, to reconvert urban wastelands and, in other cases, to preserve or even revitalise businesses that are threatened, and to defend the assets of local communities. There have been many such initiatives in Montreal, one of the North American metropolises that have been confronted with the challenge of economic reconversion for over two decades (Coffey and Shearmur 1998; Polèse and Coffey 1999; Polèse 1999; Coffey et al. 2000; Klein et al. 2003).

It is in this context of socio-economic reconversion, which particularly affects the old metropolises on the east coast of North America, that the structural changes underway in the Montreal metropolitan area for the last two decades must be understood. Although Montreal is the oldest industrial city in Canada and was Canada's leading metropolis until the 1950s, from the the post-war period, a combination of adverse economic and political factors has brought about an intensive restructuring process of its productive core.

This process was marked by industrial redeployment and the conversion to new technologies. A real 'gale of creative destruction', it has thrust upon former industrial spaces the challenging task of adaptation, a challenge that has greatly involved social actors (Klein et al. 2003). These actors undertook to develop and implement economic projects by mobilising different types of human, organisational and financial resources in order to connect their communities to the 'new economy'. Thus, social and community actors have applied social economy resources to the task

[1] This chapter is based on research funded by the OECD Local Economic and Employment Development (LEED) Programme and the Social Sciences and Humanities Research Council of Canada (SSHRC), which the authors would like to thank. The authors would also like to thank Marguerite Mendell for her comments on a preliminary version of the chapter.

of reconversion. The latter, however, envision a different reconversion, one that is implemented for the benefit of local communities rather than at their expense.

In this chapter, we will attempt to demonstrate that the active presence in Montreal of the 'civil actor',[2] that is of organisations rooted in a social movement of socio-economic reconversion, has helped to establish a model of metropolitan development that can be distinguished from other North American examples. Compared to metropolises elsewhere in North America and in Europe (Jouve 2003; Savitch and Kantor 2002), the Montreal urban regime that has been structured around the issue of industrial reconversion has certain specific characteristics. Economic growth and the development of new technologies are certainly important goals. Nevertheless, the situation in Montreal is characterised both by the important role of social movements in the urban regime and the reference point that the latter have succeeded in developing: they have made the fight for jobs the central goal of industrial reconversion policy. This is precisely what distinguishes the Montreal regime from the growth coalitions and entrepreneurial or corporatist urban regimes which have been set up in metropolises such as Boston, Detroit or even Atlanta (Elkin 1987; Stone 1989; DiGaetano and Klemanski 1999). To do this, social movements have broken new ground in their repertoire of action and have succeeded in adopting a platform of action that became a 'rallying point' for all the actors, including the public actor.

The following analysis will be presented in three parts. First, we will examine the issue of reconversion, emphasising the role of civil action in defining a territorial sphere of economic development. Second, we will refer to the opposing actors and to the social actors' specific choices to strive for a form of reconversion that associates social aims with the fight for jobs. Third, we will illustrate this type of development action using the concrete example of the actions taken to thwart plant closures in the southwest borough of Montreal, one of the marginalised districts. This borough has been the theatre of innovative experiences in local governance initiated by the civil actor in which the associational community and unions have guided the action of both public and private actors. Finally, we will underline the importance of the civil actor in establishing a plural reconversion strategy and, thus, of economic development.

Our analysis will show that the governance characteristic of the Montreal model of reconversion has less in common with current North American models than it does with what is known as the 'Barcelona model', characterised by strategic

---

[2] We prefer to use the term 'civil actor' rather than 'civil society' in this context. In our opinion, the term civil actor refers to organizations of civil society that implement projects and actions whose effects on the well-being of the communities to which they are linked, and on their governance, can be felt. We make the distinction between the civil actor (community organizations, unions), the public actor (different levels and mechanisms of the public authority) and the private actor (private capital and private enterprises). Our aim is to determine how these three types of actors interrelate in economic reconversion projects.

partnership and consensus.[3] Nevertheless, the Montreal model can be distinguished by the active presence of social movements, which have helped to shape a mode of governance in which the social actors play a key role and guide collective choices. This mode of governance strives to achieve development that is based on a pluralist economy, in which the social economy cohabits and, at times, is combined with the private economy and the public economy.

Civil Action in Reconversion of Marginalised Territories

As a result of globalisation, many economic zones of North America, Europe, Asia and Latin America have plunged into the maelstrom of deskilling and devitalisation. These are often regions that formerly lived on the exploitation of natural resources, the textile industry, wood processing, shipbuilding or the defence industry. Today, these regions must adapt to an economy that is shifting towards new technologies, creating economic turmoil with a serious social impact on these regions. A significant proportion of more vulnerable workers and populations who live in these regions are paying the price of the disinvestment provoked by the loss of their territory's economic attractiveness.

These territories, which in many cases have become marginalised in terms of development, are simply avoided by investors of the new economy. Certain zones have even become invisible in the eyes of the market and depend heavily on government measures for their survival. Others, pushed by gentrification or unbridled speculation, are turning towards more profitable sectors, often at the expense of local residents and the integrity of their environment. In several countries, however, local social actors have had enough of dependency and have chosen to take charge of their socio-economic destiny by launching innovative development initiatives.

The variety and importance of these initiatives were recently displayed at a symposium on reconversion based in civil society.[4] It was noted that they play a crucial role in the restructuring of economic spaces, even though they often result from projects that are reacting to destructuring market forces. Through these initiatives, communities, cities and regions affected by the crisis of the industrial economy connect or reconnect with the knowledge-based economy. They are responses through which local actors – in partnership with public and para-public

---

[3] In the case of Barcelona, the elaboration of development strategies has served as a forum of debate between the public institutions, social organizations and economic actors. This debate has helped to legitimise and give coherence to public and joint actions aimed at reducing inequalities and increasing the quality of life of citizens in order to democratise economic development (Borja 1998).

[4] The magnitude of the challenge faced by the territories in the continuous reconversion of their economic activities was the basis for the important symposium, held in May 2002. 'Rendezvous Montreal 2002' brought together North American, South American and European participants involved in the implementation of innovative economic and social development models (Fontan et al. 2003).

institutions, universities, unions and community organisations, as well as private and
social economy enterprises – are striving to change the market-driven course of
development of their territory and attempting to inject a social dimension into the
knowledge-based economy.

To better understand the role and place of reconversion initiatives taken by civil
society, they must be analysed in the context of relations between globalisation and
local initiative. This raises the question of new modes of regulation, new institutional
arrangements and new forms of governance that are products of the adaptation of
the local to the global level (OECD 2001a), an adaptation that should not be seen as
unidirectional.

The opening of markets and the opportunities offered by information and
communication technologies have helped to change competitive conditions and to
call into question the traditional industrial policies of government institutions. Thus,
new strategies aimed at developing the new economy and adapting production to new
market requirements have been formulated and applied. The repositioning of
national economies has often been achieved through specialisation in high-return
sectors, located in 'winning spaces' (Benko and Lipietz 1992), which has brought
about a significant shift in territorial planning and development policies (OECD
2001b).

A new framework for the growth of businesses is being set up through the
creation of continental economic zones. They have larger spaces in which to use
their mobility and flexibility to take advantage of new economies of scale and
diversified comparative advantages. Space is thus becoming a major competitive
issue. The mobility of financial capital is reshaping the supply and demand of space:
capital looks for spaces where it can accrue its investments and local actors are the
holders or suppliers of those spaces. In this context, metropolises and regional
spaces compete with each other (Klein and Fontan 2003; Pecqueur 2003).

As a result of this new competitive framework, national governments are
depending on high-return enterprises and regions. Public policies on industrial and
business development increasingly promote the concentration of activities that are
both technologically innovative and financially profitable. Finally, the leaders of
urban governance are clashing as they strive to make their spaces more attractive.
They are stimulating the development of competitive advantages through the supply
of new externalities. These measures favour the concentration of capital and
innovative services in specific zones of the large metropolises (Borja and Castells
1997).

In these new growth zones, the immediate environment of businesses and
intangible or extra-economic elements, such as the quality of networks, institutional
density, and quality of life, are highly mobilised to create comparative advantages. In
an increasingly open and globalised economy, the local level is reaffirming its capacity
to be a suitable place to create jobs and increase the competitiveness and profitability
of businesses. However, this is not the case for all local spaces, hence the renewal of
the modalities of reproduction of socio-economic inequalities. Part of the problem
of inequalities in terms of economic growth is due to the fact that the territories,
whose formerly productive sectors have declined, are not spontaneously chosen by
the most successful businesses of the new economy. Thus, piece-by-piece and very

selectively, globalisation is redrawing the map in terms of where wealth is produced and jobs are located.

Refusing the role of passive witness that the leading institutions of globalisation (Davos, G8, IMF, etc.) would like them to adopt, the actors of the communities neglected by growth are building organisations that promote local development dynamics. These organisations mobilise resources of the social and solidarity-based economy to construct local networks and harness their 'socio-territorial capital' – that is, their human, physical, organisational, cultural and identity-based resources – in their relations with economic development, while implementing new forms of collaboration with businesses, universities, unions and other social groups.

The reconversion initiatives stemming from civil society do not follow a single model. On the contrary, they reflect the breadth of reconversion and employment creation and retention activities that are currently taking place throughout the world. For researchers and practitioners, these initiatives constitute a deep well of knowledge in the field of development. The know-how demonstrated by civil society actors draws on a toolbox that dates back to the Marshallian industrial districts. This formula of the industrial district has been updated by the work of, among others, Piore, Bagnasco, Benko and Lipietz, who present the social construction of productive networks and markets as an option to neo-liberal perspectives of the transformation of Fordism (Lévesque et al. 1995; Benko et al. 1996). This approach inspires the analytical model of 'sticky places' (in a slippery space), according to Markusen's formula (1996), which 'are complex products of multiple forces: corporate strategies, industrial structures, profit cycles, state priorities, local and national politics' (Markusen 1996, p.309) The goal of local actors is to socially construct the conditions to create these 'sticky places', combining cross-regional and locally based strategies (ibid, p.310).

The Civil Actor In Strategic Choice: The Fight For Jobs In Montreal

From the point of view of civil society, marginalised territories do have resources, not the least of which is the capacity to mobilise community actors. While in the past, such mobilisation was often limited to putting pressure on the public actor or on management, reconversion experiences now rely not only on these pressures, but also increasingly on collective or social entrepreneurship to drive innovative initiatives. Underlying these experiences is a desire to prevent job losses and the devitalisation of the community. We will examine the effect of these actions in the case of Montreal.

Our hypothesis is that Montreal embodies a specific model of governance in that the traditional decision-making organisations, such as municipal structures, have been confronted over the last two decades with the pressure of new organisations whose legitimacy derives from their social representativeness. It is recognised by all actors that Montreal has been experiencing a structural crisis for several decades. The effects of this crisis, which were at first concealed by urban cosmetic operations (large urban renovation projects from 1960 to 1975, construction of the Montreal subway system in the 1960s, the 1967 World's Fair, the 1976 Olympic Games), became crystal clear to actors when in the late 1970s, there was a massive move by the industrial sectors that had traditionally supported Montreal's growth to spaces that offered more profitable conditions. This process of industrial redeployment was characterised by the relocation of businesses, job losses and a significant rise in the social assistance and unemployment rates. Indeed, between 1975 and 1992, the unemployment rate increased from 6.7 per cent to 16.7 per cent.[5]

The evidence of this crisis, which had both economic and spatial consequences, insofar as the relocation of firms and unemployment had a greater effect on districts in the urban fringe where these traditional industrial sectors were located, provoked reactions among the economic and social actors (Klein *et al.* 1998; Klein *et al.* 2003). At the risk of oversimplifying, it can be said that these reactions have given rise to three different strategies.

On the one hand, similar to the strategies applied elsewhere in North America, the representatives of the business community and of the main private and government institutions are shifting towards the development of new technology businesses and particularly towards a strategy of partnership between these businesses and the university research community. What sparked the adoption of this strategy was the task force created by the federal government in 1985 and chaired by L. Picard, a well-known figure in the university community. Based on the deliberations of its sixteen institutional leaders from Montreal, the task force produced what is known as the Picard Report, which became a reference on Montreal's economic reconversion for both public and private practitioners. The report contains a strategy that encourages private leadership, internationalisation, and the development of high-technology sectors (telecommunications, aerospace, biopharmaceuticals, information technologies and microelectronics). These objectives were to be implemented at the agglomeration level and were expected to result in planning operations at the metropolitan level.[6]

Social movements – the grassroots, or community, movement and the unions – reacted to this strategy of the business community and elites. The strategy proposed by the associational community was elaborated by organisations rooted in what we referred to earlier as the marginalised districts, that is, those put in a difficult position by industrial redeployment, first in Pointe-Saint-Charles, Centre-Sud and Hochelaga-Maisonneuve and then in all the districts of the inner city fringe.[7] These districts mobilised around the community leaders to defend their assets and undertake development strategies that are adapted to new economic conditions, and to respect the interests of the local population. The main results of this mobilisation can be seen in what is described by the actors as a 'community economic development' strategy and in the creation of organisations of Montreal's Community Economic Development Corporations (CEDCs) devoted to the application of this intervention strategy (Fontan 1991; Hamel 1991).

---

[5] The situation has improved since then. From 1995 onwards, the unemployment rate has gradually decreased, reaching 8.4 per cent in 2002. However, this rate has remained higher than that of Toronto (7.4 per cent) and the Canadian average (7.7 per cent) (see Statistics Canada, CANSIM, table 282-0053 and Catalogue no 71-001-PIB).

[6] See 'Report of the Advisory Committee to the Ministerial Committee on the Development of the Montreal Region'.

[7] Which are called 'peri-central districts'. See Klein et al. 1998.

The central goal of the CEDCs is to promote the partnership of the actors in their districts. Their aim is to get actors to work together and to implement partnership-based development projects, allowing actors to make contact with each other, and to identify common goals. The second main goal of the CEDCs is to support local entrepreneurship in order to help create local jobs. Finally, the third goal is to enhance the employability of the jobless, that is, to provide individuals with the skills needed to re-enter a job market undergoing intense restructuring. The districts or boroughs constitute the space in which the CEDCs act. In this way, their existence as intermediary public spaces allows the potential of local territories as a framework for collective action rooted in social movements to be realised. This constitutes an important change in community action, one that has been the subject of numerous debates even within the Montreal social movement.

RESO (the *Regroupement pour la relance économique et sociale du Sud-Ouest de Montréal*) is an example of such a CEDC. RESO has taken on the challenge of reconversion through direct intervention. When plant closures multiplied, when poverty was spiralling upwards, and when there were no resources to attract firms, RESO showed that it was possible to change this trend, improve the employment situation within existing businesses, achieve social insertion and training of the jobless, attract firms, revitalise the existing infrastructure, and create jobs and wealth. Thus, there was a shift from reactive, confrontational action to pro-active entrepreneurial action. Since their creation, the CEDCs have maintained a key role in promoting concertation at the local level and local development. The Quebec government's recent decision to make RESO a Local Development Centre gives recognition to this role.[8]

The third type of strategy, which like the preceding type also originates in social movements, is the strategy resulting from union action. Since the early 1980s, unions have adopted a strategy that has re-oriented their action and transformed them into important development actors. In reaction to globalisation and industrial redeployment, the unions have focused their action on the fight for jobs by creating investment funds and tools to prevent plant closures.

A good example of this strategy is the creation of retirement funds for the purpose of fighting business closures and investing in job creation. The first and largest fund of this type was the *Fonds de solidarité* (Lévesque 2000), created in 1983 by the *Fédération des travailleurs et travailleuses du Québec* (FTQ, Quebec Federation of Labour) with the explicit goal of creating jobs. The FTQ fund currently has 500,000 shareholders and assets of nearly five billion Canadian dollars. This then prompted Quebec's second-largest labour confederation, the *Confédération des syndicats nationaux* (CSN, Confederation of National Trade Unions), to create a retirement fund in 1996, appropriately called *Fondaction*, with goals similar to those of the FTQ fund. However, this new fund was more oriented towards venture capital investment in social economy enterprises. *Fondaction* has a quarter of a billion dollars and nearly 50,000 shareholders.

---

[8] Local Development Centres were created and funded by the provincial government to give local initiatives a support. They were composed by local civil society representatives.

Again, from the perspective of the fight for jobs, the unions have established forms of action that seek to anticipate crises in firms before they arise and to suggest changes that could help prevent crises. This is true, for example, of the FTQ organisation, *Urgence-Emploi*, and of the CSN's employment watch project, the *Projet de veille pour l'emploi*. These services support the efforts of local unions to prevent massive layoffs. They develop, jointly with management and government organisations, recovery plans for firms having difficulties, thus suggesting a determination on the part of the unions to participate in entrepreneurial governance.

The creation of CEDCs by the community association movement, and the establishment of union tools and services devoted to supporting the creation or consolidation of jobs and to preventing business closures, are among the actions carried out by the Montreal social movement to make the fight for jobs one of the strategic orientations of Montreal's reconversion. In so doing, both the unions, with their significant financial assets and the community organisations with their potential for social mobilisation, have become actors recognised by both government authorities and the business community. Moreover, this focus on employment provides direction to the otherwise uneven actions of Montreal actors as a whole.

The Case of South-Western Montreal: Partnership-Based Governance of an Economic Space

The territory of South-Western Montreal includes the former municipalities of Lachine, LaSalle and Verdun, all of which became boroughs of the new City of Montreal on 1 January 2002. It also includes the southwest district of the former City of Montreal, which is itself made up of six districts that belong to the oldest population areas of Montreal Island. Thanks to the Lachine Canal, South-Western Montreal became a bastion of industrialisation in Canada. Several steel and manufacturing plants were set up on its banks in the first half of the 19th century. These enterprises employed thousands of workers who lived close to their place of work.

When the Lachine Canal was closed to shipping in the 1960s, South-Western Montreal entered a long period of economic decline. One by one, the plants either closed or laid off a very great number of workers. As a result of a drop in the birth rate and the population exodus to more affluent neighbourhoods, the population began to decline. Public investment mainly took the form of welfare transfers: social assistance or employment assistance. Private investment decreased. Investors, put off by the territory's less attractive image, either fled or bypassed it.

Origins of Community Economic Development in South-Western Montreal

In the late 1960s, under the leadership of social actors, often clergy members, the population was mobilised and reacted to the socio-economic deterioration of their territory. Between 1965 and 1980, many important initiatives were undertaken by Montreal community groups – citizens' committees, community clinic and pharmacy, legal aid clinic, day care centres, housing committees, food banks, community newspapers, and local roundtable.

In 1984, community groups in Pointe-Saint-Charles broke new ground by creating an organisation dedicated to the economic development of their territory, the *Programme économique de Pointe-Saint-Charles* (PEP). This was Montreal's first CEDC (there are twelve today). In 1988, faced with the continuing collapse of the territory's economy, local actors came together to form the *Comité pour la relance de l'économie et de l'emploi du Sud-Ouest de Montréal* (CREESOM), which developed an action plan whose central proposal was to extend PEP's mission to the entire district. In 1989, PEP became the *Regroupement pour la relance économique et sociale du Sud-Ouest* (RESO). Less than three years later (1992), the CEDC *Transaction pour l'emploi* was created in the territories of LaSalle and Lachine. Its mission was similar to that of RESO, that is, to create and retain jobs, basic and occupational training, employment insertion and partnership of community actors to revitalise the local economy. The initiatives of these two organisations, which quickly became key resources in the region, were responsible for the creation of collective strategies for revitalising firms, particularly through local alliances for employment. These strategies were implemented in close co-operation with the unions.

Revitalisation of the Dominion Bridge Plant

The fight to prevent the closure of Dominion Bridge, a major plant in the district, provides a good illustration of that part of the work carried out by South-western Montreal's CEDCs. This plant, established in the district in 1879, was one of the last vestiges of Montreal's early steel industries. Throughout its history, the firm greatly contributed to the development of the area, providing work to many residents of South-Western Montreal. When the plant closed in 1998, it employed 250 people.

The *Métallos* (steelworkers union) local section 2843 (affiliated to the FTQ) had been present in this plant since the early 1950s and many of the employees who still worked at 'la Dominion' had more than 25 years of service. Throughout these years, a distinctive culture of work relations had been established. This culture was built on the employer's respect for the workers' skills, responsibility and expertise. It constituted the main comparative advantage of the firm, a manufacturer of custom-built parts. Similarly, for many of these workers immersed in a world where individuals are defined in relation to their work and employer, the employer had taken on the status of 'prudent administrator', to use an expression which has been codified in Quebec law.[9] Thus, many workers came to identify themselves almost entirely with their work and their struggles as a result of many union battles.

In spring 1994, Cedar Group Inc., a U.S. company, bought the Dominion Bridge plant. Shortly after the buy-out, both the shareholders and union members learned that the plant had been mismanaged. Conflicts erupted over appropriate solutions, even among the shareholders. Supporters of an the approach based on 'profitability through improvement of production capacity' squared off against those in favour of a purely speculative approach based on 'management of financial assets'. Made aware

---

[9] A union member interviewed in February 2002 used these terms in reference to respect and formalism.

of the tensions, the union turned to the FTQ's *Urgence-emploi* services to assess the firm's financial situation.   The diagnosis revealed the seriousness of the firm's financial situation.   The union therefore asked the FTQ *Fonds de solidarité* to determine whether the firm could be revitalised.

The shareholders in favour of the speculative approach took temporary control of leadership within the Cedar Group.   In April 1996, Dominion Bridge Corporation, as the Cedar Group renamed it, acquired the MIL-Davie shipyard in Lévis, owned by the Government of Quebec and worth over $100 million, for the nominal sum of one dollar.   In the autumn of the same year, the Lachine plant laid off approximately 100 workers.   The local union feared that the jobs and contracts at the Lachine plant would be transferred to Lévis.   Convinced that bankruptcy was imminent, the entire FTQ structure was mobilised and, in December 1996, its president appealed publicly for the federal and provincial governments to get involved.

Two years later, in April 1998, yet another attempt was made to alert the population to the plant's situation.   The local union called on *Transaction pour l'emploi* to implement a rescue operation for the firm.   The *Comité de survie et de relance* (survival and revitalisation committee) was formed, which included the following actors:

- Steelworkers union, local section 2843 (an FTQ affiliate).
- FTQ's *Conseil régional du Montréal métropolitain* and *Urgence-Emploi*.
- *Transaction pour l'emploi* CEDC.
- *Société de développement des artères commerciales de Lachine* (SIDAC Lachine).
- Lachine Chamber of Commerce and *Regroupement Affaires Lachine Inc.* (RALI).
- Marguerite-Bourgeois School Board and CÉGEP André-Laurendeau.
- Municipality of Lachine.
- Members of the provincial National Assembly and federal Parliament
- Local organisation of the Parti Québécois.

In September 1998, the company declared bankruptcy.   Efforts to save the bankrupt firm were made simultaneously at several levels.   While the unions, members of the public, and community groups – sometimes supported by Lachine municipal employees – took turns picketing outside the plant to prevent the order book and equipment from being removed, political pressure was applied in an effort to find a solution to the crisis.   At the same time, the FTQ *Fonds de solidarité* sought prospective buyers for the plant.   The highest levels of the steelworkers union and FTQ organisations were mobilised.   Press conferences were organised to raise awareness among the Quebec population and demonstrations were held to publicise the situation at the local level.   The local alliance for the revitalisation of Dominion Bridge put considerable stock in local residents' identification with the century-old plant.

After taking many different actions (preventing a trustee in bankruptcy from getting back contracts, blocking the Mercier Bridge, many demonstrations and press conferences, etc.), the *Fonds de solidarité* managed to interest a Montreal medium sized-firm, the Groupe Au Dragon Forgé Inc. (ADF Group Inc.), in buying out Dominion Bridge.   An agreement was reached between the two financial stakeholders: ADF would acquire 51 per cent of Dominion Bridge shares while the *Fonds de solidarité* would hold 49 per cent, plus a 20 per cent participation in the ADF Group. The plant was rebaptized ADF Heavy Industries Inc. and resumed operations in November 1998, nearly three months after it had gone bankrupt and mobilisation to save it had begun.

As a condition of ADF's involvement, when the plant was re-opened, the local union had to sign a collective agreement with its new employer. The employees had to make concessions, particularly as regards work organisation (reorganisation of job descriptions). However, the gains sacrificed were largely recovered over the three-year life of the agreement and, in particular, about one hundred workers regained their jobs when the plant re-opened. Subsequently, the number of jobs gradually increased, reaching its pre-closure level.[10]

Local and Participatory Governance for a Plural Economy?

As shown in the case of Montreal and, in particular, in the case of the Dominion Bridge rescue in South-western Montreal, the fact that employment is a rallying point for actors has an effect on the type of development that is implemented. The public actor, the private actor, and the civil actor each have a role in partnership-based experiences. In this context, the interventions of the civil actor can help to attract investments and develop entrepreneurship in conditions that are profitable for both the local community and outside private capital.

The objectives of these experiences differ, but they are all shaped by the same strategic goal: employment. This goal also makes it possible to mobilise market resources, non-market government resources and non-monetary resources from citizens, which adapts the sense of citizen participation to the context posed by a fragmented territoriality. The pursuit of these objectives – and the difficult tradeoffs that this implies – cannot be achieved without a democratic approach that gives a significant role to debate and to the exchange of information. Economic restructuring thereby takes on a new look, more diversified than restructuring that is uniquely focussed on the goals of production. It calls upon new economic sectors and new actors. It also requires that the political means and mechanisms of supporting such initiatives be renewed. We see in this approach the germ of a new model of development that constitutes an alternative to the 'one best way' view of growth, committed only to an unrestrained free market.

The industrial reconversion initiatives springing from civil society come within what is called the 'social economy' in Quebec and the 'solidarity-based economy' in France. But the following should be stressed: these initiatives renew the means of action traditionally used by the organisations that claim to draw on them. Because

---

[10] They continued to increase until the events of 11 September 2001, in the United States. Since most of the firm's production consisted of large structures for construction projects in North America, the post-9/11 economic crisis in the United States delayed certain contracts, which weakened the firm and gave rise to job cuts.

they are neither the product of the spontaneous adjustment of the market nor of the intervention of the government, these initiatives are by definition part of the third sector. Nonetheless, the successful or ongoing reconversions have encouraged the creation not only of social enterprises but also of private firms, without excluding public sector initiatives. In experimenting with strategies that balance economic and social development, will these reconversion initiatives lead to or result in bridges being built between the various types of firms and organisations, reflecting the development of a 'plural economy'?

That is what we hope for in the case of Montreal. In Montreal, the involvement of the civil actor is contributing to the redefinition of the concept of 'economic reconversion' by broadening the understanding that public and private actors currently have those types of activities to be reconverted. This opens other avenues for tomorrow's economy, which will undoubtedly be based more and more on relations and knowledge (Borja and Castells 1997). Seen from this angle, economic reconversion can play a role in fostering high technology activities as well as in the development of culture and services that improve the quality of life of citizens. In addition, the knowledge developed through the initiatives of civil society has the advantage of combining already well-proven sectoral development strategies with innovative strategies for supporting entrepreneurship. A particular facet of this knowledge is that it is taking shape in innovative sectors of intervention, while reconnecting the linkages between the local level and the wider web of world developments.

Thus, although the new economy is propelling us into the field of technological innovations, these developments are also intimately related to social innovations. On this level, reconversion initiatives sponsored by civil society in Montreal are particularly innovative because they construct new interrelationships between the development actors, that is, public, private and civil actors. These bridges make it possible to envisage a plural form of governance and economy that are likely to be responsive to the needs expressed by the population, including those of territories facing difficulties while, at the same time, being compatible with the elites' aspirations for growth.

References

Benko, G., M. Dunford and A. Lipietz (1996) Les districts industriels revisités. In B. Pecqueur (ed) *Dynamiques territoriales et mutations économiques.* Paris, L'Harmattan, pp.119-136.

Benko, G. and Lipietz, A. (ed) (1992) *Les régions qui gagnent, districts et réseaux: les nouveaux paradigmes de la géographie économique.* Paris, PUF, 1992.

Borja, J. (1998) Ciudadania y espacio publico, *Reforma y democracia*, No. 12. Electronic journal of 'Centro latinoamericano de administración para el desarrollo CLAD,' Venezuela. Document downloaded. (www.clad.org.ve/reforma.html)

Borja, J. and M. Castells (1997) *Local & Global: Management of Cities in the Information Age.* London, Earthscan Publications.

Canada (1986) *Report of the Advisory Committee to the Ministerial Committee on the Development of the Montreal Region.* Ottawa, Minister of Supply and Services of Canada.

Castells, M. (1997) *The Power of Identity.* Cornwall, Blackwell.

Coffey, W., C. Manzagol and R. Shearmur (2000) L'évolution spatiale de l'emploi dans la région métropolitaine de Montréal, 1981-1996, *Cahiers de Géographie du Québec*, Vol.44, No.123, pp.325-339.

Coffey, W. and R. Shearmur (1998) Employment Growth and Structural Change in the Urban Canada, 1971-1991, *Review of Urban and Regional Development Studies*, Vol.10, No.1, pp.60-88.

DiGaetano, A. and J.S. Klemanski (1999) *Power and City Governance*. Minneapolis, University of Minnesota Press.

Elkin, S.L. (1987) *City and regime in the American republic*. Chicago, University of Chicago Press.

Fontan, J.-M. (1991) *Les corporations de développement économique communautaire montréalaises. Du développement économique communautaire au développement local de l'économie*. Doctoral thesis, Department de Sociology, Université de Montréal, Montréal.

Fontan, J.-M., J.-L. Klein and D.-G. Tremblay (2001) Les mouvements sociaux dans le développement local à Montréal: deux cas de reconversion industrielle, *Géographie Économie Société*, Vol.3, No.2, pp.247-280.

Fontan, J.-M., J.-L. Klein and B. Lévesque (2003) *Reconversion économique et développement territorial: le rôle de la société civile*. Sainte-Foy, Presses de l'Université du Québec.

Fontan, J.-M., J.-L. Klein and D.-G. Tremblay (eds) (1999) *Entre la métropolisation et le village global*. Sainte-Foy, Presses de l'Université du Québec.

Fontan, J.-M. and J.-L. Klein (2000) Mouvement syndical et mobilisation pour l'emploi: renouvellement des enjeux et des modalités d'action, *Politique et société*, Vol.19, No.1, pp.79-102.

Hamel, P. (1991) *Action collective et démocratie locale. Les mouvements urbains Montréalais*. Montréal, Presses de l'Université de Montréal.

Jouve, B. (2003) *La gouvernance urbaine en questions*. Paris, Elsevier.

Klein, J.-L., J.-M. Fontan, D.-G. Tremblay and D. Bordeleau (1998) Les quartiers péri-centraux: Le milieu communautaire dans la reconversion économique. In C. Manzagol and C. Bryant (eds) *Montréal 2001*. Montréal, Presses de l'Université de Montréal, pp.241-254.

Klein, J.-L., D.-G. Tremblay and J.-M. Fontan (2003) Systèmes locaux et réseaux productifs dans la reconversion économique: le cas de Montréal. *Géographie Économie Société*, Vol.5, No.1, pp.59-75.

Klein, J.-L. and J.-M. Fontan (2003) Reconversion économique et initiative locale: l'effet structurant des actions collectives. In J.-M. Fontan, J.-L. Klein and B. Lévesque (eds) *Reconversion économique et développement territorial: le rôle de la société civile*. Sainte-Foy, Presses de l'Université du Québec.

Lévesque, B. (2000) *Le Fonds de solidarité FTQ: un cas exemplaire de nouvelle gouvernance*. Université du Québec à Montréal, CRISES.

Lévesque, B., J.-M. Fontan and J.-L. Klein (1995) *Systèmes locaux de production: Réflexion-synthèse sur les nouvelles modalités de développement régional/local*. Montreal, Université du Québec à Montréal, Cahiers du CRISES, No.9601.

Markusen, A. (1996) Sticky Places in Slippery Space: A Typologie of Industrial Districts, *Economic Geography*, 72(3), pp.293-313.

OECD (2001a) *Local Partnerships for Better Governance*. Paris, OECD.

OECD (2001b) *Territorial Outlook*. Paris, OECD.

Pecqueur, B. (2003) Construction d'une offre territoriale attractive et durable: vers une mutation des rapports entreprise/territoire. In J.-M. Fontan, J.-L. Klein and B. Lévesque. (eds) *Reconversion économique et développement territorial: le rôle de la société civile*. Sainte-Foy, Presses de l'Université du Québec.

Polèse, M. (1999) La dynamique spatiale des activités économiques au Québec: Analyse pour la période de 1971-1991 fondée sur un découpage centre-périphérie, *Cahiers de géographie du Québec*, Vol.43, No.118, pp.1-24.

Polèse, M. and W. Coffey (1999) A District Metropolis for a District Society? The Economic Restructuring of Montreal in the Canadian Context, *Canadian Journal of Regional Science*, Vol.XXII, Nos.1 and 2, pp.23-40.

Sassen, S. (ed) (2002) *Global Networks, Linked Cities*. London, Routledge.

Savitch, H. and P. Kantor (2002) *Cities in the International Marketplace. The Political Economy of Urban Development in North America and Western Europe.* Princeton, Princeton University Press.

Stone, C.S. (1989) *Regime Politics: Governing Atlanta (1946-1988).* Lawrence, Kansas University Press.

Chapter 10

Putting Locally Based Management to the Test in French Public Policy

Philippe Warin

Introduction

In most political discourse today locally based management is viewed as an essential feature of effective and democratic public action, and a feature of good governance principles determined at an international level. It is seen both as defining the questions to solve, and presenting the solutions required, within every conceivable area of state intervention.

Yet the principle of proximity is hardly anything new. Throughout history it has been rediscovered time and again that the distance between governors and governed is essentially what characterizes political regimes (totalitarianism being the extreme form of overwhelming proximity). Nevertheless, over time issues evolve and the subjects of debate change. The main issue is no longer the need for local democracy to transform a political-administrative system marked by limited debates on local issues, as it was thirty years ago. It is now the necessity for other modes of production of public policies, closer to the inhabitants of the territories concerned, with a view to increasing the effectiveness and transparency, and consequently the legitimacy, of public policies. It is for this reason the principle of proximity is currently so successful. It encompasses everything from social action to security, health, transport, urban planning, consumption etc., as well as more recent and emblematic issues such as sustainable development or the social and solidarity economy. Since achieving pragmatic solutions and transparent results has become the golden rule for good public management, the dominant discourse is the same everywhere; that is to act in proximity.

This chapter does not argue for political, locally based management as a way of countering the invasion of the social world by the market economy (Warin 2002a). Rather, we focus on a series of observations that challenge the general celebration of the proximity principle in France. We argue that locally based management, apparently so advisable, is not achieved satisfactorily via processes of transformation of public policies, despite their attempt – in different respects and for different reasons – to more or less directly make it effective. This is the paradox that we consider here, taking into account three processes constituting some of the pieces of the puzzle (i.e. a public policy model whose final image is still uncertain): the contractualization of public policies; partnerships between public services and associations; and individualization of service delivery.   The contradictory or problematic nature of each of these processes is then highlighted.   Our hypothesis is that while the issue of locally based management can be apprehended from different angles, its implementation through processes of transformation of public policies raises a multitude of questions that need to be considered before embarking on any debate on its democratic value.   In other words, we believe that the probability of locally based management being a lever for democracy depends primarily on trends in the processes of contractualization, partnership and individualization of public policies, and on their consequences.   From this perspective the proximity issue can be seen above all as a 'macro issue'.

Our intention is, however, not to appear over-critical or disillusioned.   The application of the proximity principle is far from being an accomplished fact.   We could say that it is a constantly evolving process.   Most governments have clearly understood that they have to move closer to the people and develop action in the field that meets the population's concrete needs.   In this respect, since the shock of the 2002 presidential elections when the populist far right discourse was concretised on a massive scale through the ballot, the current French government has adopted a rhetoric with plebeian tones in an attempt to be heard 'from below'.   But application of the proximity principle is also a solution from a managerial point of view.   For a start, proximity can be a convenient tactic for states subjected to European regulations and budgetary directives.   In a sense this amounts to localizing *public* action in order to optimise state resources by transferring certain expenditure to local and regional authorities.[1]   The latter nevertheless highlights their dedication to locally based service delivery, ensuring their own control of a public policy (e.g. *départements* with regard to social action and the regions regarding training and employment).[2]   Since proximity suits everyone, we can see why it is so successful and, in particular, why public decision-makers ask for more.

The Contractualization of Public Policies

We would like to turn away from the recurrent debate on the reign of generalized negotiation that runs through multiple networks of actors and professionalised forums

---

[1] It is partly the surpluses of local administrations (and especially that of central administrative bodies, the ODAC: the *Caisse d'amortissement de la dette sociale*, the *Fonds de réserve des retraites*, the *Fonds de réserve de la Couverture Maladie Universelle*, etc.) that limits the public deficit of the state and social security. For instance, at his 2 September 2002 press conference, the minister of national education expressed the wish for local authorities to participate in financing educational jobs performed by supervisors and auxiliary educators, and thus develop new *'proximity supervisory jobs'*. In his framework document the prime minister provided for a budget for *'possible cofinancing with local authorities of a new system of supervision in schools'* (*Le Monde*, 5 September 2002, p.6).

[2] In the system of French administrative institutions, the *Conseil Général* and the *Conseil Régional* represent political institutions at two territorial levels: the Department (100 in all – 4 of which are overseas); and the Region (22), each with their own executive and administration. Depending on the competencies attributed to these levels of authority and on the political options of these institutions, both the Departments and the Regions are sometimes key public policy actors.

and refers strategic decisions back to closed arenas. These important aspects have been thoroughly analysed elsewhere (see, in particular, Gaudin 1996, 1999). Their study has made it possible to account more precisely for the systems of actors and interaction involved in policy production, as well as the tools used in the different policy phases to coordinate the diversity of interests, competences and resources. This has spawned an approach to the notion of governance that we could call 'positivist'. The idea is that the multiplication of arenas of negotiation, the generalization of contracts and the development of regulatory agencies constitute three key modalities of 'new public action'. Despite the significance of these studies, once we have concluded that none of this facilitates either the actors' control or the citizens' involvement, we see that it is of little use in our discussion.

On the other hand, if we bear in mind this transformation of public action, we note that the public services concerned by the implementation of contractual policies (devolved state services or those of local/regional authorities) are confused by the multiple aspects relative to content and processes. Once bound by their partnership they are faced with incompatibilities between programmes. Very often they have different definitions of the targeted publics, the territories concerned and the duration of actions (Fontaine 1999; but this is not specific to France, see Diamond 2002). Moreover – and this is often the biggest problem – they have different ways of functioning when it comes to drawing up budgets and spending funds, making the financial monitoring of programmes and the comparison of results something of an acrobatic exercise (Leroy 2000). Meetings between representatives of the state and of a regional council, for example, show that it is not generalized negotiation that prevails throughout but misunderstanding. Everyone spends a great deal of time trying to understand the other party's programme, and equivalences, yet the aim of contractual policies is to distinguish between different programmes. In the evaluation stage of a regional urban policy (incorporated into the state-region plan), we noted that the main recommendations concerned the need to homogenize the territories of intervention, the criteria on which aid was granted and the heads of expenditure. These recommendations stemmed from the fact that it was impossible to compare any of the various partners' actions, and that the policy in question was far from being a shared reality that could easily be evaluated. From experience, this often is the main result of evaluations of contractual policies.[3]

For these reasons, the public and private sector actors involved in the implementation of the same policy often follow action programmes that do not necessarily have the same content. From this point of view and considering their complexity, contractual policies intended to solve targeted problems seem insufficiently prepared in the programming phase of their cycle prior to their implementation (Knoepfel, Larrue and Varone 2001). Consequently, the aim of managing social problems directly, via a series of programmed actions, can be compromised from the programming stage by unresolved differences between the rationales, rules and so on, of each institution.

---

[3] For seven years we sat on the Rhône-Alpes regional scientific committee for evaluation and in 1999 undertook a joint comparative study of evaluation in the Regions for the Rhône-Alpes agency for human and social sciences.

These shortcomings in programming identified in the implementation stage of contractual policies impact negatively on the possibility of locally based management. The aim of managing social problems *in situ* remains a vague general intention. Recommendations to involve the concerned groups are therefore seldom taken into account. Since programme content lacks consistency or coherence, proximity with populations and territories in programme implementation does not function as a category of operational action. That is why some actors, having experienced the difficulties of implementing it comment: 'everyone's talking about locally based management, participation, etc., but they all interpret it in their own way, without really knowing how to go about it'. Consequently, instead of being perceived as an operational objective, locally based management is seen as an ill-identified principle of action or simply a watchword. In any case, it by no means constitutes a system of shared codes and rules enabling the actors to identify with a collective strategy. This ultimately affects the actors' behaviour, hence the caution and minimal commitment by decision-makers and implementers to the collective processes of steering and evaluation of contractual policies (see Fontaine and Warin 2001). In other words, locally based management is confronted with a problem of assent whenever there is a lack of rigour in the programming of the contractual policies within which it is inscribed.

The underlying problem is that area-specific contractual policies (like others) opt for 'steering by objectives' if not 'by projects'[4] to avoid cumbersome procedures of the various partner institutions, for partnerships have to be boosted. In these conditions, locally based management seems basically to be an objective to achieve, or an ambition to share, without any further definition. For the actors of implementation, more used to 'steering by procedures' (especially public actors), this is far too imprecise.

This constraint is not felt everywhere to an equal degree. When the programming of policies involves actors who also participate in implementation, the processes of diagnosis and the preparation of objectives relate more directly to issues in the field. The fact that the experience of professionals and institutions can play a part often makes it possible to come closer to the problems actually encountered by concerned groups and to the practical conditions of action. This conclusion can realistically be drawn from several programmes of social and urban rehabilitation of so-called 'difficult' neighbourhoods. The explanation for this lies in the quality of

---

[4] Three types of steering of policies or organizations can be distinguished: 'steering by procedures' aims to translate objectives into rules and procedures, to require the actors of implementation to follow the rules and procedures and to check whether their behaviour is consistent with them; 'steering by objectives' exists in translating ambitions into objectives, in motivating the actors involved and in regularly adjusting action to bring results into line with objectives; 'steering by projects' aims to make ambitions common to all the actors concerned by the implementation of the policy, programme or action, and to maintain their mobilization around the implementation of the project (Bauer and Laval, 2002).

relations between actors and in their engagement in projects.[5] These conditions of success are also noted in experiments with participative management at the local level (Diamond et al. 1997; Rosenberg and Carrel 2002). They may vary, depending on whether the local councillors are prepared to play the card of locally based management or not (Faure, Gerbaux and Muller 1989), and on the clarity and soundness of the policy rationales. The asymmetry of situations also stems from conditions of eligibility for public financing. When grants are subject to the creation of devices for dialogue and participation, in which the actors responsible for actions and sometimes the target groups are involved (as in France with the first urban policy programmes – they contribute substantially to incorporating objectives of proximity).

Partnerships Between the Voluntary Sector and Public Services

Partnerships with voluntary sector organisations are an important mode of production of social policies. Cooperation between associations and public authorities is recognized today, and is both officialised and normalized. This observation confirms the fact, often noted in recent years, that public authorities can no longer produce policies alone. In the domains of social and health services, the voluntary sector has been an unavoidable partner for a long time. On many issues and projects it is the voluntary sector that makes proposals. Recently, the examples of minimum income support (RMI – *Revenu Minimum d'Insertion*), universal medical aid coverage (CMU – *Couverture Maladie Universelle*) and the struggle against exclusion show the decisive role of large national associations. These organizations take advantage of their constant presence in many political and administrative networks, and employ particularly effective strategies to recruit experts and ensure media coverage (Warin 2002b). This also applies to health where relations of interdependence between public authorities and associations, for instance, have been instrumental in organizing the fight against Aids (Pinell 2000).

As regards outputs, the partnership with the voluntary sector has progressed considerably. In fifteen years there has been a shift in various areas (health, social and urban policy, employment, culture, etc.), from a policy of subsidising voluntary sector activities to one in which relations have been contractualised and targeted actions or projects are financed through the purchase of services. A form of public action based on partnerships between public services and associations at the service of locally based management of various social problems has thus been developed.

This development is partly explained by the importance that partnerships with associations now have in the production of public policies. While the tradition of articulation between 'state welfare' and 'private charity' previously caused the boundaries between acts of solidarity and public intervention to become somewhat fuzzy (especially in the medical-social sector), the current situation of disengagement

---

[5] Rather than citing books on the subject, of which a large number exist, we refer the reader to an address that could be useful to those interested in a complete bibliography as well as some conclusions on various French experiments. This is the address of the *Centre de ressources et d'échanges pour le développement social et urbain – Rhône-Alpes* in Lyon: crdu@crdu.org or crdsu.secretariat@free.fr.

of the State seems to reinforce a trend towards instrumentalisation and interdependence between public services and private initiatives (whether for-profit or not). The need to mobilize such extensive human resources in the production of public policies, and to individualize answers, makes it necessary to think in terms of partnerships so as to spread the costs, associate complementary competencies, and to deploy forms of action as close as possible to the concerned groups.

The success of certain themes, especially the struggle against social exclusion, frees space for the expression of the 'living forces' that wish to define themselves outside the framework of regular competencies concerning social assistance and action. This trend is not specifically French (Le Galès and Négrier 2000) and is also found in other European countries and in many OECD countries. In fact it is often even more pronounced elsewhere, where the concept of welfare is understood to be both a matter of a welfare mix and of welfare pluralism (Evers 1997). Yet, despite the wish for consensus implicit in partnerships, they have mainly been imposed by the State rather than being free, negotiated and balanced. It seems that in this game the State, as donor and regulator, keeps the upper hand (Damon 2001). To stimulate and frame their partnerships with voluntary associations, public authorities have taken a series of important political initiatives in recent years, combined with legal and statistic normalisation measures (Warin 2002a).

Whereas partnerships with the voluntary sector are based on a logic of locally based management, in some ways they show that there remain grey areas and that major obstacles can appear.

Grey Areas

This is precisely what is revealed by the hushed but crucially important debate launched by the Jospin government in circles close to the ministry in charge of social welfare and solidarity, on the need to recognize the essentially non-monetary economic value of associative activity (or, more broadly, not-for-profit private-sector activity). For these actors, the wealth produced is essentially human and social (prevention, resocialisation, involvement, solidarity), and therefore public financing of associations and more generally of the social welfare and solidarity sector is seen as a necessary investment from an economic, social and democratic point of view. With regard to finance, the perception is not the same because the issues of public grants and taxation have to be addressed in order to comply with pressing European directives. In other words, the political choice was made in 1999 (will it be adhered to by the current government?) to explicitly favour locally based management by the public authorities by encouraging and normalizing partnerships with associations. Yet there has been no answer to the main political question of whether this locally based management founded on the voluntary sector is perceived above all as a financial cost, in which case it is reduced or stopped because considered too expensive, or as an investment, in which case it can be reinforced as a mode of public action. This was a key issue among the parties of the left in power. It also is at the heart of current debate on growth and wealth (Orléan 1999; Généreux 2001).

Intense uncertainty prevails today as to whether the development of public locally based management via partnerships with the voluntary sector ought to be continued. However, the direction this locally based management through partnerships should take remains unanswered and is unlikely to be resolved until the role granted to the involvement of the voluntary sector has been clarified. Is this type of management merely curative (in which case the voluntary sector is simply a relay for public services) or is it more preventive, implying the involvement of concerned groups (which would mean incorporating a philosophy of association into public policy)? We thus see that in order to develop, the idea of locally based management relies upon the political treatment reserved for policy actors. The institutionalisation and normalization processes of partnerships with associations nourish the idea of public locally based management, but with neither clarity as to the goals nor certainty as to their pursuit. The only certainty is that proximity rhetoric makes it possible to justify the use of an additional actor required by public services are overburdened in many respects. A concern to economise on expenditure has for the moment at least ranked higher than the social and economic advantages of proximity with residents.

Major Obstacles

Nothing has been gained, in real terms, for at the same time the logic of partnerships with the voluntary sector is hardly well received by those primarily concerned, i.e. local public services (of the State or local/regional authorities) (ISM-CORUM 2001). Local services use the themes of versatility and proximity to avoid the introduction of collective services that involve the voluntary sector in their production. The reason for this is straightforward: the fear that a contractual actor will rob the statutory authority of its prerogatives and resources to the point that the authority itself may be dismantled. This is the key problem of public locally based management running counter to the interests of public service agents and bodies.

These grey areas and obstacles show how locally based management, as a model of intervention through partnerships, remains largely subordinate to the wider debate on the future of public services. Consequently, the institutionalisation of locally based management, in this partnership form, now seems something for the distant future. As regards the desire for a locally based management that allows the creation of new welfare relations (that is closer to the people and more democratic) the moment when the voluntary sector works harmoniously with local public services is still far off. Yet it is this close working relationship that will allow such local public services to express their conception of proximity, based on the involvement and empowerment of the groups concerned.

Individualizing Service Delivery

The process to be considered now presents itself in several ways simultaneously. It examines the changes under way in public policies and administration. These changes are intended to improve service delivery, yet they produce unexpected effects that complicate or compromise, if not contradict, that objective. In the following section we discuss a number of observations on administrative modernization and, to conclude, on public policies. Note, however, that these comments warrant further development and that other examples could be taken into account.

Public Administration

By definition so-called 'administrative modernization' policies are supposed to facilitate the move to administrations that are modern because suited to users' needs (this is the core of the 'new public management' doctrine prevailing in all OECD countries). These modern administrations are believed to be the perfect example of locally based management because they are close to the public they serve. Yet things seem less simple when we examine proposals on both sides, for instance, the public agents and the users.

Observation 1: the management of public institutions For the past few decades, government institutions have been subjected to successive structural adjustments to policies of varying degrees of severity, depending on the country. During the 1990s, in France, successive devices constructed a model of devolved management of objectives, means and results, with the stated aim of empowering services and agents.[6] A logic of locally based services is strongly present in this model since the aim (regarding the empowerment of services and agents) is to increase the performance of the administration of contact with users.

The paradox is that this model subordinates agents to performance criteria that they apply only very partially while, at the same time, it totally ignores the fact that in the course of their work those same agents practice proximity in their relations with users. In many instances these practices enable them to achieve the type of performance that is so essential for the public service yet absent from the criteria; treating users equitably (Warin 2002c). In other words, closeness to users, as advocated by management, appears to have little in common with the daily practice of agents who are in direct contact with users. Is the rationale so different that it produces disinterest in these agents to such an extent? This seems particularly strange in so far as the approach in terms of knowledge management (Prax 2000) has been incorporated into the techniques of public administration management – although there is a good chance that the way in which services and administrations actually work is more prosaic than the descriptions in management textbooks.

The problem is not so much the fact that management performance is difficult to evaluate, even in countries that are soundly committed to this option (Boyne 2001), but that this administrative model, grounded in the principle of an increasing engagement by services and agents, disregards the actual, existing engagement of agents confronted with both management objectives and user pressure. For the agents, contact with service users is a daily reality. The questions, for them are: what is the public service rationale of these reforms; what it means in terms of missions; and in relation to that, what are its advantages? This indicates that the objective of

---

[6] Initially these were devices ('centres of responsibility' stemming from 'service projects') that compared the activities and expenditures of services on the basis of the then popular 'means/objectives plans'. They were followed by devices ('service contracts', 'quality charters', 'service engagements') that preferred the logic of management by results – which could serve to distinguish between productive and unproductive services – rather than resource management which had become obsolete.

increasing local contact within policies whose aim is to modernize administrative practice faces a problem of implementation because, in many respects, it seems too restrictive and too distant from the realities experienced by agents in direct contact with users.

Observation 2: users' rights In many areas claims-rights are developing as a constant (and superfluous) revision of human and citizens' rights. This apparently limitless extension of 'rights to ...', which grants recognized advantages and prerogatives to individuals, contributes towards a watering down of 'modern rights' in which the legal is conceived in terms of the individual only (Villey 1983; Goyard-Fabre 1999). Claims-rights therefore tend to be substituted for existing freedom-rights recognized not for individuals in particular, but for the population in general.

In public administration we are witnessing a similar inflation of claims-rights (Warin 2002d), although the extension and consolidation of civil rights granted to users have important consequences. At the formal, administrative procedure level junior officials are increasingly bound by a multitude of obligations in terms of service quality (deadlines, explanations, motivations for decisions, etc.). That is why, to avoid being constrained by conventions that are hardly realistic in relation to their real working conditions, a majority of them choose to focus primarily on content; that is, on actual results in terms of services delivered. They concentrate essentially on the implementation of administrative regulations governing access to social rights, rather than to civil rights that give an impression of citizenship.

In this respect, the development of claims-rights granted to users by law is intended to bring the public service closer to its public. However, in concrete terms, it compels officials to refer primarily to impersonal rules. Dealing with users' requests at the point of contact, rather than being simply a matter of following new conventions, leads to a far more consistent re-examination of regulatory systems. Where locally based management is seen as a way of trimming administrative processes, it seems to reactivate legalistic forms of administrative functioning. Consequently, there is uncertainty today on exactly which logic of locally based management to pursue.

Public Policies

This trend in which the public service is brought closer to users is accompanied by a tendency for policies to differentiate the provision of public services on the basis of an increasingly individualized approach to requests. Multiple areas are concerned. In a sense, there is an awareness of the diversity of users and needs, and consequently an attempt to adjust service delivery accordingly.

Without entering into the details of the processes of adjusting service delivery, we wish to note some paradoxes in the cases mentioned above of the police and judiciary and of schools (see Box below).

Individualized treatment of social demands in all directions

The police and judiciary have been experimenting with proximity for a long time. There is no need to go into the details here of all the devices created in these sectors in the past two decades; what is important is the fact that they have all been 'local'.[7] The need for proximity in the treatment of requests is even becoming the justification in the race for partnerships. For instance, problems of violence in schools was taken into account in policies at the same time as new security policies (Dumoulin and Froment 2003), more or less linked to urban policy (*politique de la ville*), were decided (Donzelot and Wyvekens 1998). Today a penal apparatus is being established that also provides assistance for victims in schools, on the basis of close partnerships between local actors of national education, the police and the judiciary. This means that in addressing such problems, the authorities have seized the opportunity to present the addition of cross-cutting devices to other cross-cutting devices as a necessity. Eventually, by building up networks of actors characterized by the closeness of the different members, confusion sets in among the partnerships and the object of the action is swamped.

In a more recent example, on the agendas of European governments, policies to combat exclusion activate the proximity principle to the same degree. Devices are created that not only target their publics but move about on territories in order to – or so it is claimed – be accessible and reactive, and hence effective. Education is one of the main areas in this respect. The struggle against the crucial problem of absenteeism is developing at national and local level, and on a European scale. Devices are created to counter the problem of 'school drop-outs' (e.g. the national programme 'New chances' and the European pilot project 'Second-chance schools'). Local contact is sought at all levels of government.

The paradox of mediation devices   Specialists on the police and judiciary have shown how the need to produce results and to avoid or reduce congestion (due to the huge numbers of pending files) has led to the creation of locally based facilities grounded in a logic of mediation and conciliation. It was primarily because of the administration's need to rank and sort cases that this solution was developed. The paradox is that instead of facilitating relations with victims and the treatment of their cases, these devices exclude part of the targeted public, reducing the take-up of police and judicial services.

The increase in the number of complaints reported has resulted in larger numbers of cases being referred to alternative modes of conflict resolution, such as social or penal mediation. However, these negotiated modes are perceived partly as a type of

---

[7] *Conseils communaux de prévention de la délinquance* (CCPD), *Groupes locaux de traitement de la délinquance* (GLTD), *Plans locaux de sécurité* (PLS) and, more recently, *Contrats locaux de sécurité* (CLS), *Maisons de la Justice et du Droit, Conciliateurs locaux*, etc., as well as the countless *Observatoires* (research institutes).

second-class justice, causing the possible plaintiffs to refuse to talk. They cannot understand the information campaigns designed to persuade citizens to report certain offences of which they were victims. Devices set up concerning violence in schools, for example, tend to increase the take-up level. But here again, to avoid an excess of reported cases, the institutional partners agree on selection criteria that, in turn, result in non-take-up by some potential applicants. Hence, the paradox also relates to the fact that the devices introduced to deal with the consequences of necessarily selective administrative functioning themselves produce a form of selection. The quest for proximity can consequently result in a phenomenon of disinterest within public services, and possibly in non-take-up by targeted groups.

The paradox of devices to fight against exclusion As in many other sectors, policies to combat exclusion exist in education. The idea is clearly to remedy the problem of early exit from schooling. The paradox is that relay devices created for the pupils concerned make their return to the regular education system difficult. In turn, the high failure rate of those who resume ordinary schooling has the end result of even greater marginalisation of the concerned groups. In this case, the extension of schooling in these devices, called for by many specialists, can be counter-productive accentuating segregation and unequal opportunities.

This example shows that the educational and social consequences of the locally based management proposed to reintegrate difficult pupils can be very uncertain. It also shows the difficulties of treating certain problems in the absence of preventive policies'upstream'. In France that absence is obvious in most sectors. It is precisely because a whole range of social problems have not been perceived in time and are not always dealt with by means of preventive measures (acceptance of their costs is a virtually insurmountable problem for the public authorities). There is an attempt now to make up for those shortcomings via proximity devices. The question now is whether it is sufficient, and up to what point it will seem preferable, to spend money on curative proximity rather than investing in preventive proximity. The success or failure of locally based management depends both on the objectives that are set and the resources granted initially. This shows that locally based management alone is not an adequate answer, in any case certainly not in the social domain in which it is so sought after.

The limits of the project logic Finally, we wish to point out that the *project logic* at the heart of many public policies today (education, training and professional insertion, the struggle against unemployment, the struggle against poverty and social exclusion, etc.) does not facilitate achievement of the locally based management objective that it is supposed to serve. It is a proximity logic *par excellence* since it tends to place individuals back at the heart of institutional action. The idea is to involve individuals and to free the potential of each one by calling for collaboration or individual engagement in a project. In this respect, we can but wonder about possible correspondence of the proximity theme with that of 'social capital' (Putnam 2000). This seems more appropriate than the links made with currently trendy notions such as civic literacy (Milner 2002) that is focused far more on a question of participation, in so far as the social policies hardly correspond to the characteristic of 'reflexive policies' (Giddens 1994) aimed at turning individuals into real actors, producers of public events and not simply beneficiaries. The proposed personalized trajectories of integration are even claimed to aim to replace the request for participation by a strategy of involvement, with a view to avoiding possible conflicts that could challenge the distribution of power (Donzelot 1991). We might add that another advantage is to afford a convenient explanation in case of failure, in so far as the result is connected to individuals' efforts. In any event, one thing seems fairly sure. Project logic aims to involve and empower individuals, but not with an aim of establishing a system of roles and rules based on a conception of participative democracy. Far from it. Individuals are simply encouraged to take their fate into their own hands.

Considered from the angle of the project logic, the idea of locally based management replaces the 'individual problems/collective responsibility/public response' sequence by an 'individual problems/public answer/individual responsibility' sequence. The former constituted the framework of a society that considered solidarity to be the principle and the objective of public policies. The latter expresses the reconstitution of social interaction in an individualistic mode, as opposed to the abstraction of universalistic rights that the public authorities are having difficulty guaranteeing today (Castel 1995). In this quest for proximity there is an illusion at play, that is, the project may be a palliative for everything. The project logic changes nothing of the structural inequalities that are often the root of the social problems against which it is supposed to act. Proximity thus sought target publics – assuming that they can be reached[8] – therefore seems more like a dogma than a solution based on rigorous analysis of the facts.

Conclusion

To sum up, we see that locally based management as an objective, and even as a practice already at work, is largely subordinate to a set of factors related to the transformation of public policies, the outcome of which is essentially still unknown. With this rapid survey we have seen that locally based management is contingent on policy objectives, programme content, management style, the adhesion of the actors of implementation, and the ability to recognize and incorporate the proximity logics already at play (with the different cases of the associations and reception staff). Behind each element we see that locally based management depends on political choices that, in the main, have still not been made. What are the missions that we want to assign to public services? Are we ready to develop preventive policies? Can we provide the human resources required to develop individualized policies effectively? Do we have the capacity and the will to adopt modes of involving groups that are not strictly administrative? Do we know how to avoid the stumbling block of pre-constructed modalities; steering by objectives, contractualization and projects?

The processes of transformation of public policies considered here indicate that locally based management may be considered as a framework for action, but that it is

---

[8] The growing awareness in France, following that of other countries, that the non-take-up of welfare and public services highlights the illusion of individualization and contractualization of the public offer as an effective solution when insufficient human resources are deployed.

far from corresponding to clearly defined and mastered realities. That is why, before theorizing on its democratic advantages in general, it seems advisable and urgent to examine in depth the processes underway that define its conditions of existence both directly and indirectly. In any case, locally based management that is neither totalitarian nor destructive of personal liberty is, *a priori*, always good for democracy, provided that we understand how it is constructed. In this respect, the examination presented here reports somewhat reserved observations requiring more in-depth analysis:

- Contractual policies, that can be an important mode of development of locally based management, are insufficiently prepared to impact favourably on the behaviours of actors of their implementation.
- Partnerships between the voluntary and public service sectors, which are also one of the possible modes of locally based management, are far from facilitating the dynamic involvement of the concerned groups; at this stage they encounter the fears of public services who are on the defensive.
- Individualization of the public offer, whether at the level of administrative functioning or public policies, either ignores existing proximity logics (those of public agents) or recreates logics opposed to proximity (mediation devices that again produce a selection of target groups and push some towards non-take-up; devices to combat exclusion that in turn produce exclusion; etc.), or sometimes even undoes solidarities and excludes the eventuality of a participative approach (the project logic that individualizes solutions and responsibilities).

Considered in relation to these processes, locally based management seems to be a phenomenon with multiple forms and uncertain effects. To know how it can serve as a lever for participative democracy, we need to examine in what way these processes are themselves participatory. In light of what we already know of the processes of transformation affecting public policies, one observation is clear: apart from processes of involvement of civil society currently developing in France through a series of procedures applied to the infrastructure and health sectors (Vallement 2001; Rui, Ollivier-Trigalo and Fourniau 2001) or scientific and technological choices and risks (Callon, Lascoumes and Barthe 2001), for the rest (i.e. essentially contractual policies, social policies and reform of the administration, where explicit reference is made to the need for proximity with territories and populations), we are still waiting for the application of the 'deliberative principle' (Blondiaux and Sintomer 2002) or for a return of participation that associates citizens more and produces political innovations comparable to those developed in other countries (McLaverty 2002). In many cases the idea of integrating consultative procedures still seems unfeasible, whereas in other countries (e.g. the Netherlands) the consultative model has already been outmoded by a deliberative model (Klijn and Koppenjan 2000). In France public administration and public policies constitute, *a priori*, ideal places for developing forms of locally based management although they are still largely closed to experimentation in new forms of citizenship.

We can hardly believe that this is because reform processes in France are not firm enough (Rouban 1998). Nevertheless, it is in those countries where drastic political choices have sometimes been made (e.g. the UK), and where structural reforms regarding decentralization and privatisation with major consequences for public services have been undertaken, that forms of locally based management that have been sought-after and applied have most clearly created areas of participation, although often with mediocre results (Diamond 2002).

References

Bauer, M. and C. Laval (2002) Management Public, Management Privé, *Informations Sociales*, No.101, pp.12-21.

Blondiaux, L. and Y. Sintomer (2002) L'impératif délibératif, *Politix*, No.57, pp.7-35.

Boyne, G. (2001) Planning, Performance and Public services, *Public Administration*, Vol.79, No.1, pp.73-88.

Callon, M., P. Lascoumes and Y. Barthe (2001) *Agir dans un monde incertain*, Paris, Seuil.

Castel, R. (1995) *Les métamorphoses de la question sociale*, Paris, Fayard.

Damon, J. (2001) La dictature du partenariat, *Informations Sociales*, No.95, pp.36-49.

Diamond, J. (2002) Decentralization : New Forms of Public Participation or New Form of Managerialism. In P. McLaverty (ed) *Public Participation and Innovations in Community Governance*, Burlington, Ashgate, pp.123-140.

Diamond, J. *et al.* (1997) Lessons from Performance Review, *Local Government Policy Making*, No.23, pp.27-30.

Donzelot, J. (1991) *Face à l'exclusion. Le modèle français*, Paris, Editions Esprit.

Donzelot, J. and A. Wyvekens (1998) *La politique judiciaire de la ville : de la 'prévention' au 'traitement'. Les groupes locaux de traitement de la délinquance.* Research report, Paris, Mission de recherche Droit et justice/IHESI/DIV.

Dumoulin, L. and J.-C. Froment (2003) L'école et la sécurité : les politiques de lutte contre la violence scolaire. In J.-C. Froment, J.-J. Gleizal and M. Kaluszynski (eds) *L'État à l'épreuve de la sécurité intérieure. Enjeux théoriques et politiques*, Grenoble, PUG, à paraître.

Evers, A. (1997) Le tiers secteur au regard d'une conception pluraliste de la protection sociale. In J.-N. Chopart *et al.* (eds) *Produire les solidarités. La part des associations*, Paris, Imprimerie nationale, pp.51-60.

Faure, A., F. Gerbaux and P. Muller (1989) *Les entrepreneurs ruraux*, Paris, L'Harmattan.

Fontaine, J. (1999) Quels débats sur l'action publique? Les usages de l'évaluation des politiques publiques territorialisées. In F. Bastien and E. Neveu (eds) *Espaces publics mosaïques*, Rennes, PUR, pp.285-305.

Fontaine, J. and P. Warin (2001) Evaluation in France, a Component of Territorial Public Action, *Public Administration*, Vol.79, No.2, pp.361-381.

Gadrey, J. (2000) *Nouvelle économie, nouveau mythe ?*, Paris, Flammarion.

Gaudin, J.P. (ed) (1996) *La négociation des politiques contractuelles*, Paris, L'Harmattan.

Gaudin, J.P. (1999) *Gouverner par contrat. L'action publique en question*, Paris, Presses de Sciences Po.

Généreux, J. (2001) Manifeste pour l'économie humaine, *Esprit*, No.7, pp.141-171.

Giddens, A. (1994) *Les conséquences de la modernité*, Paris, L'Harmattan.

Goyard-Fabre, S. (1999) *L'État. Figure moderne de la politique*, Paris, Armand Colin.

ISM-CORUM (2001) *La proximité, quels enjeux pour les services publics?*, Villeurbanne, Mario Bella Edition.

Klijn, E.H. and J.F.M. Koppenjan (2000) Politicians and Interactive Policy Making: Institutional Spoilsports or Playmakers, *Public Administration*, Vol.78, No.2, pp.365-388.

Knoepfel, P., C. Larrue and F. Varone (2001) *Analyse et pilotage des politiques publiques*, Bâle/Francfort, Hlebing and Lichtenhahn.

Le Galès, P. and E. Négrier (2000) Partenariats contre l'exclusion : quelles spécificités sud-européennes?, *Pôle Sud*, No.12, pp.3-12.

Leroy, M. (2000) *La logique financière de l'action publique conventionnelle dans le contrat de plan État-Région*, Paris, L'Harmattan.

McLaverty, P. (ed) (2002) *Public Participation and Innovations in Community Governance*, Burlington, Ashgate.

Milner, H. (2002) *Civic Literacy. How Informed Citizens Make Democracy Work*, Plymouth, University Press of New England.

Orléan, A. (1999) *Le Pouvoir de la finance*, Paris, Odile Jacob.

Pinell, P. (2000) *Une épidémie politique : la lutte contre le sida en France de 1981 à 1996*, Paris, PUF.

Prax, J.Y. (2000) *Le guide du knowledge management. Concepts et principes du management de la connaissance*, Paris, Dunod.

Putnam, R. (2000) *Bowling Alone. The Collapse and Revival of American Community*, New-York/London, Simon and Schuster.

Rosenberg, S. and M. Carrel (2002) *Face à l'insécurité sociale. Désamorcer les conflits entre usagers et agents des services publics*, Paris, La Découverte.

Rouban, L. (1998) Les États occidentaux d'une gouvernementalité à l'autre, *Critique Internationale*, No.1, pp.131-149.

Rui, S., M. Ollivier-Trigalo and J.-M. Fourniau (2001) *Évaluer, débattre ou négocier l'utilité publique?*, Paris, Rapport INRETS, No.240.

Vallemont, S. (2001) *Le débat public : une réforme dans l'État*, Paris, LGDJ.

Villey, M. (1983) *Le droit et les droits de l'homme*, Paris, PUF.

Warin, P. (2002a) Les associations en France : les enjeux politiques d'une reconnaissance juridique et économique, *Pyramides*, No.6, pp.65-82.

Warin, P. (2002b) The Role of Nonprofit Associations in Combatting Social Exclusion in France, *Public Administration and Development*, No.22, pp.73-82.

Warin, P. (2002c) *Les dépanneurs de justice. Les 'petits fonctionnaires' entre qualité et équité*, Paris, LGDJ.

Warin, P. (2002d) Les droits-créances aux usagers: rhapsodie de la réforme administrative, *Droit et Société*, No.51-52, pp.437-453.

Chapter 11

Participating in Development: The Case of a Public Open Space in Villeurbanne in the Lyon Conurbation

Jean-Yves Toussaint, Sophie Vareilles, Marcus Zepf, Monique Zimmermann
Translation by Philip Booth

The configuration of actors in urban development projects has been transformed as a result of recent changes in French public policy. The proposals for improving public open spaces in Lyon are particularly revealing of the capacity of these networks to be open to, and closed off from, a wider public, a capacity that is more and more bound up with the means of communication. Both local authorities and national organisations, in particular central government, have encouraged partnership in urban development projects, in the knowledge that such encouragement would be generally welcomed and receive the support of technical officers, including those who are actually responsible for the implementation of projects. Indeed, new ways of carrying out development projects has favoured the emergence of new actors, in particular public relations agencies and mediators. All this poses a number of questions about the nature of this concern for involving the end-users of these urban improvements, about the actual effects of the change for their involvement, and about the way in which projects are carried out.

These questions are dealt with in this chapter in the context of a project for improving the Place Lazare-Goujon in the centre of Villeurbanne, one of the *communes* of the Lyon conurbation. This project represents the fruit of more than ten years of experimentation in public participation in the creation and improvement of public spaces and was aimed at capitalising on that experience. It is for this reason a point of reference for the way in which professionals within greater Lyon currently practise public involvement.

This study has been carried out as a piece of action-research with the cooperation of the public authorities involved, the *commune* of Villeurbanne and the *communauté urbaine* of Lyon, who allowed the research team to attend working sessions as well as public meetings.

An Introduction to the Context of Lyon

Since the creation of the *communauté urbaine* of Lyon on the 1 January 1969, the conurbation has been characterised by a nesting of units of authority that are both political and administrative. The *communauté urbaine* of Lyon, one of nine such structures created in the 1960s to give some coherence to the largest of the French provincial conurbations, has been rebaptised Greater Lyon (*le Grand Lyon*) and comprises 55 *communes* (the base unit of local authority in France) of which Villeurbanne is one. Three institutions constitute the core of Greater Lyon: a council for the *communauté*; an executive office; and a president. The council is made up of 155 members drawn from each of the municipal councils in the *communes* and determines policy in all the areas for which Greater Lyon is responsible.[1] The executive office is made up of vice-presidents, each of whom has a particular portfolio and is answerable to the president. The president and the vice-presidents are elected by the mayors of the *communes* and of the *arrondissements* of the city of Lyon and by the different municipal elected representatives.

This system of political representation at the level of the conurbation gives some actors within the system omnipresence. Members of the council of Greater Lyon are at one and the same time representatives of the whole conurbation and, as municipal councillors, representative of their own electorate. In Villeurbanne, for example, the mayor is also the third vice-president of Greater Lyon.

From 1989 onwards, there has been an evolution in the practice of implementing urban projects and the basic premise on which it was taken has been renewed. Development policy has centred on public open space that has been seen as the means of recreating the city and enhancing the quality of its urban design. This policy was based on three principles: restoring the primacy of political leadership in urban development; taking public space as representative of social organisation and interaction; and insuring public 'ownership' of public space [2] (Toussaint and Zimmermann 2001, Toussaint 2003b).

In practice, the treatment of public spaces has been approached in recognition of the complexity of the uses to which they are put and of the different elements of the population that use them. They are seen as places available to residents for particular activities and for the acting out of daily life. To achieve that end, the technical services of Greater Lyon have been reorganised: an 'open spaces' department, cutting across traditional departmental boundaries, acts as the 'client' for work on public open space, a strict distinction has been made between the client department and the contractors, and more generally, a culture of open spaces has been established within the technical services as a whole.

---

[1] Greater Lyon is responsible for development, town planning, housing, economic development, the road network, water services, collection and disposal of household waste etc. Green spaces and street lighting remain the responsibility of the *communes*.
[2] In this way, political discourse has emphasised the general over the particular and the global over the local. Public policy for open space has aimed at influencing social behaviour and resisting the functional privatisation of open space, especially in the management of road networks and car parking.

Over the period of some ten years, about 300 improvement projects have been completed within the conurbation (Azéma 2001). These projects have often incorporated underground parking. Greater Lyon, with its particular responsibility for parking provision, has devolved responsibility for the construction and management of these parking spaces to a *société d'économie mixte* (a public-private joint venture company), Lyon Parc Auto (LPA), as a result of which it has become a key player in the improvement of public open spaces. The transformation of the practice of urban improvement, quite apart from improving the quality of open space, has also been intended to draw in new actors, beginning with designers and artists, and then, through various forms of participation, incorporating both residents and users.

It is this involvement of residents and users, which has been given little formal recognition,[3] that poses the biggest questions for both the commissioning and the implementation of such projects. To this extent, public participation is the subject of discussion and experiment in the public authorities, which are trying to find a way of going about it. In Greater Lyon the departments of Social Development[4] and Public Open Spaces have gradually been developing participation in a systematic way, through social surveys, partnerships, and public meetings, exhibitions, and the creation of *maisons de quartier* (neighbourhood centres).

The 1995 elections gave Villeurbanne a new mayor. A new deputy mayor for local democracy and citizenship was appointed and he put in place neighbourhood councils[5] from 1996. Such provision is to be seen in the light of Villeurbanne's history[6] as a socialist local authority, alongside the residents' assemblies of the 1930s, the neighbourhood committees and newspapers from the beginning of the 1970s, and the consultative committees with neighbourhood organisations at the end that decade (Meuret 1982).

From 2000, discussion on participation in the Lyon conurbation has centred on developing a culture of participation and communication with the public. To this end, several initiatives have been put into effect, notably the creation of a 'participation watchdog', which is still undergoing development. This unit, which will exist within the planning agency, will evaluate the experience of participation in Lyon, and appears to have as its source Greater Lyon's mission to facilitate participation and the new organisational structure, which is being created to achieve that end. In effect, there is a move to integrate all forms of participation in a kind of federal structure the better to control the wide range of individual initiatives.

---

[3] Of some 30 mechanisms for involving the public in projects concerned with public open space in Greater Lyon, only four have been given legal recognition: initial consultation, neighbourhood councils, the *conseils d'initiative et de consultation par arrondissement* (arrondissement councils) and the *enquête publique* (public inquiry).
[4] This department has responsibility for urban policy implementation in the large public housing estates of the Lyon conurbation.
[5] Neighbourhood councils, of which there are seven in the town, consist of about 30 people, (local volunteers with an elected representative as chair) and are given the task of 'developing and overseeing projects, and of encouraging local citizen involvement in the neighbourhood through debate and by disseminating information'(Municipal information leaflet, Villeurbanne)
[6] And also more generally in the French tradition in public involvement (Ragon 1977, Mollet 1981, Blanc 1995, Callon, Lascoumes and Barthe 2001).

To start this global strategy for participation, a group of projects was selected: the project for the banks of the River Rhône and the Place Lazare-Goujon at Villeurbanne. The list has since been extended. The Place Lazare-Goujon was chosen also because of its experimental aspect, which included the research element, represented by this chapter. The project involved the *commune* of Villeurbanne, Greater Lyon and the Agence d'Urbanisme de Lyon (the planning agency for the conurbation). The experiment in the case of the Place Lazare-Goujon can also be seen in a wider context of projects aimed at achieving more democratic procedures for urban development set out in two charters: the charter for outdoor space (Villeurbanne) and the charter for public participation (Greater Lyon). To some extent, the Place Lazare-Goujon was used as a testing ground for the debates being conducted within Greater Lyon. The project cannot, therefore, really be described as typical of the range of approaches to participation adopted, but does bear witness to the evolution of practice, its obstacles, the brakes on its effectiveness, and its achievements. The project is thus a test of the validity of the methods that Greater Lyon and its *communes* are attempting to adopt.

The Origins of the Place Lazare-Goujon Project

Place Lazare-Goujon is situated between the town hall and the Théâtre National Populaire (TNP). This part of the town centre was built in the 1930s and is subject to heritage protection by the Ministry of Culture. Specifically, the area was designated as a *zone de protection du patrimoine architectural, urbain et paysager* (ZPPAUP; conservation area) in 1993. As has often been the case, work to the square was the result of building an underground car park, which created an opportunity for improvement above ground. Indeed, there were actually three projects involved in the Place Lazare-Goujon: a car-park project, an open space improvement project, and a participation 'project'. Organisationally, this rather unusual set-up meant that there were three project leaders: one for the underground car park, one for the open space project and one to run the public participation. Indeed, for the purposes of description we need to distinguish four separate elements within the overall project:

- The car park project, which concerned the development of the underground car parking area whose overall cost was approximately €11 million;
- The public open space project, which involved the re-design of the place Lazare-Goujon at a cost which varied between €1.4 and €1.7 million;
- The participation project, whose objective was to develop an experimental approach to public involvement, and for which the cost has never been evaluated;
- The place Lazare-Goujon project as whole, which was the combination of the three preceding projects.

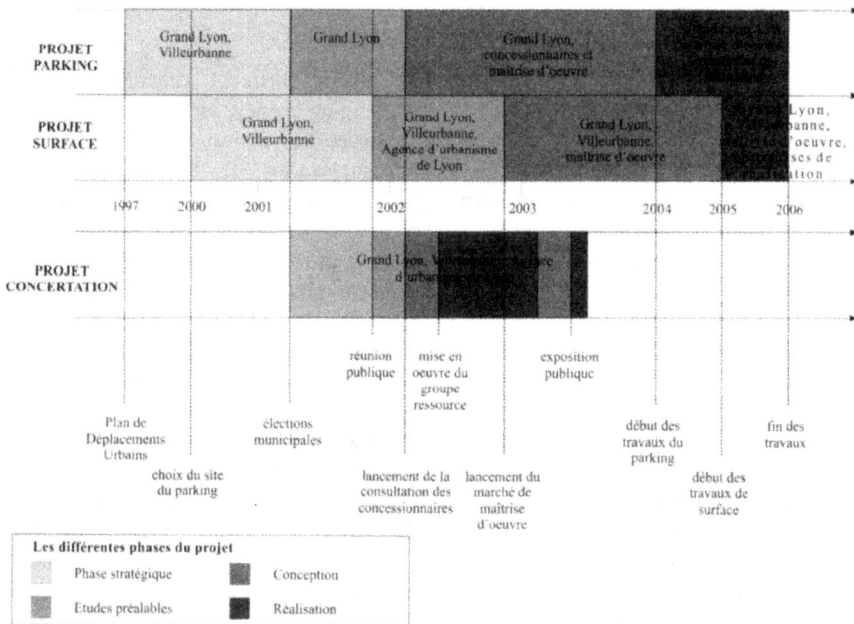

Figure 11.1 The Place Lazare-Goujon Project Programme

The Car Park

The car park project involved three organisations: the *commune* of Villeurbanne (its mayor, its deputy mayor with responsibility for public transport and highways, and its director of environment department); Greater Lyon (the vice-president for transportation policy; and the transportation department) and the car park operator.

The urban transport plan adopted by Greater Lyon in January 1997, made allowance for the provision of a car park in Villeurbanne town centre. In 2000, place Lazare-Goujon was identified as a possible site for such a car park. In March 2001, the place Lazare-Goujon project became part of the manifesto of the current mayor in the municipal election. Following the election, the project was the subject of two debates, and a feasibility study was undertaken. This brought to a close the first, strategic, phase of the project.[7] Several months later, in November, the car park project was presented at a public meeting: the capacity of the car park, the number of levels, the form of construction, the points of access for pedestrians and vehicles, were all identified. Preliminary investigations continued. Discussions were held with potential operators, and then an operator (Lyon Parc Auto) was selected. Work was scheduled to begin in 2004.

---

[7] It is possible to identify four phases in the project: a strategic phase; preliminary investigation; detailed design; and implementation (Toussaint, 2003a). The move from one phase to the next marked a certain irreversibility: decisions were taken and choices made (Midler, 1998)

168   *Metropolitan Democracies*

The Public Open Space

The open space project brought together several different organisations: the commune of Villeurbanne, Greater Lyon, the planning agency, local amenity groups, residents' and users' groups, associations representing commercial interests and employees, and other public and private sector organisations. To these must be added the National Movement for the Improvement of Housing (Mouvement national pour l'Amélioration de l'Habitat), a team of researchers, Architectes des Bâtiments de France (ABF; the national historic buildings agency) and a local television station.

The decision to redesign the open space was taken at the same time as the decision to construct the underground car park. The demolition involved in creating the car park, together with the 'poor condition'[8] of the square, lay behind the decision to proceed with the redesign (local councillor, public meeting, 28 November 2001). Indeed, the future of the square was broached publicly at the same time as the car park project, at the time of the local election and at a public meeting in 2001. From this moment onwards, the main elements of the project were fixed: the project would 'free the open space of cars and by so doing improve the convivial quality of the square...' (local councillor, public meeting 28 November 2001). In 2002, the first studies were conducted – a sociological study and one carried out by the ABF – and a project programme developed. The programme was announced publicly in October 2002. The work to the Place Lazare-Goujon, located within a ZPPAUP, would have to accord with regulations for the zone and at the same time was intended to '...modernise its appearance... [and] ...respect the area's identity' (technical officer, resource group meeting, 1 October 2002). With these considerations in mind, the existing street furniture would be retained, including the pools, pergolas and benches, but modifying the fountains and the street lighting would be possible. The project team would need to include a conceptual designer and an artist (resource group meeting 1 October 2002). Following the presentation of the programme, the design phase of the project could begin. Greater Lyon and the commune of Villeurbanne opted for a competition and selected four designers who were invited to submit proposals in March 2003. The winner, the In Situ group from Lyon, was announced in July 2003. From then until January 2004, the designs were worked up to the point at which construction could start. Completion is expected at the end of 2005.

Public Participation

Developing an experimental approach to public participation for the place Lazare-Goujon project involved three organisations who were concerned with developing procedures that could be used in the conurbation as a whole, namely Greater Lyon, the planning agency, and the commune of Villeurbanne. This set-up involved professional planners and, in particular, experts in participation (notably the sociology research unit) and the public in the guise of representatives of residents' and users' associations, local retailers, the unions, private sector enterprise, a local television station and other local media, and individual participants.

---

[8] Quotations here and in the rest of this chapter are taken from verbatim accounts of meetings observed by Sarah Russeil and Sophie Vareilles.

Figure 11.2 View of Place Lazare-Goujon

This participation project dates from the amalgamation of the three public authorities at the end of 2001. The commune of Villeurbanne was hoping to assemble a group of local inhabitants and users to assist in the development of the programme for the treatment of the public open space in order to improve its current state. At the same moment, Greater Lyon asked the planning agency to review the procedures for public participation. The planning agency, aware of the project underway in Villeurbanne, favoured using the project as a pilot for setting up a participation resources centre that they would lead on Greater Lyon's behalf (Working paper, Agence de l'Urbanisme, 2001).

At the beginning of 2002, discussions were organised among the three principal organisations in order to define working methods and the financing of the project (internal meeting, 13 February 2002). The project was finalised in May 2002 with the setting up of a working group. At the end of May, the group's objectives were developed. It was to be a 'resources group',[9] in order to emphasise its role in spreading its experience with participation among the town's inhabitants. From May to July 2002, the resources group was involved in the development of the public open space project, which was presented in October 2002. The number of meetings originally envisaged was reduced (resources group meeting 16 May 2002). These arrangements were evaluated at a meeting at the beginning of 2003. In March 2003, the four design groups taking part in the competition for the public open space presented their projects to the resources group, ten days before the official closing date for the project schemes.

Forms of Participation in the Place Lazare-Goujon Project

From the point of the research, public participation in the context of the redesign of the place Lazare-Goujon includes the full range of methods:

- The experiment in public participation for Greater Lyon;
- The work of the resources group, including all its meetings;
- The traditional arrangements for consultation required by the law, in effect the public meetings and the exhibition of proposals.

A Public Meeting

The first public meeting, held on 28 November 2001 at Villeurbanne town hall, was a good example of this type of consultation. It brought together elected representatives, including the mayor of Villeurbanne, from the commune and Greater Lyon, technical officers, including the director of the open spaces department, and about 150 members of the public, made up of inhabitants and representatives from groups and political parties. This meeting was effectively the starting point for public participation in the place Lazare-Goujon project. It was, in the words of an elected representative 'an information and consultation meeting about the underground car park and the restoration of the surface of the square ...'

This meeting consisted of three elements: a political presentation, a technical explanation of the project, and a discussion. Councillors were committed to the political presentation of the project as a means of setting the scene for the discussion. The mayor recalled in his introduction that '...the place Lazare-Goujon project was part of my programme at the local election ... there was opposition during the election, but today there is no question of wondering whether the project will go ahead, only how it will be done ...'. At the same time, councillors defined the major elements of the project and created a framework for the development by identifying the space to be freed of cars. In so doing, councillors also fixed the limits of

---

[9] It was a designation by default; there were already other working groups within Villeurbanne. By using this title, elected representatives and technical officers wanted to avoid any confusion with older designations (facilitator, working group meeting, 23 May 2002).

discussion and declared certain elements as non-negotiable because they arose from earlier decision making. The political agenda had effectively made the decision to proceed with the project irreversible in order to limit what might then be possible to 'improve'.

The technical presentation consisted of a slide show of 29 slides. It reviewed the history of the project and outlined its characteristics: 400 car parking spaces, three levels of parking; the points of access for cars, and the pedestrian access within the theatre. This number of places was justified by the findings of a survey that was intended to establish the needs of each of the groups involved with the car park. By the same stroke, the technical officers proceeded to rationalise the irreversible decisions taken on the square's development.

Discussion, as such, only began after the technical presentation and included elected representatives and some twenty members of the public; technical officers remained silent. Almost half of those members of the public who spoke were representatives of groups or political parties. The comments were mainly about the place Lazare-Goujon project as a whole. Some of the comments had as much to do with the role of the speaker, and many of the arguments were well known to everybody. The rare comments made by speakers outside the institutional framework mainly concerned practical problems of parking, the number of spaces to be reserved for residents, the effect of changes to the highway network on traffic in nearby streets, and in those parts of the network that were modified, and the cost of, and the need for, such modifications.

At the end of the meeting no dates for future meetings were fixed and no intention to minute the proceedings was announced.

The Resources Group

The resources group was the key structure within the participation project: it was set up in May 2002 was intended to be operative until the end of the project, which was fixed for the beginning of 2006. Its life span was thus fixed for a three-year period. To start with, the group was tasked with 'filling out [through its discussions] the brief for the redesign of the square' (elected representative, resources group meeting 16 May 2002). In addition to its involvement in the development of the project, it also saw its objective as 'taking its work outside the group itself' (elected representative, resources group meeting 16 May 2002).

The resources group brought together the principal actors in the place Lazare-Goujon project: local councillors including the mayor of Villeurbanne; local authority officers from Greater Lyon, Villeurbanne and the planning agency; members of neighbourhood committees and unions; representatives of other public and private sector agencies, including the TNP, the post office, the municipal police force, three members of local television, one mediator, two researchers, and from time to time, external participants. This group was connected to the professional, political and even personal networks of the actors involved in the project;[10] with its members being co-

---

[10] For example, the director of local television, the mediator and the researchers (who were members of the group) belonged to a circle of acquaintances of the leader of the participation project at the planning agency.

opted. Beyond this co-option, the composition of the group was the subject of debate between the head of the participation project and elected representatives. The debate focused on how enough space could be found for those that the elected representatives and the local authority officers called 'residents'. To that end, 'we tried to make sure that the residents were not crushed by the others ... so we reduced the number of people who were not residents ...' (mediator, working group meeting 6 May 2002). The researchers noted that, during meetings, the residents group was made up of members of neighbourhood committees, of users' groups such as those representing pedestrians and the paralysed as well as of representatives of the post office, the municipal police or the theatre. These 'residents' were characterised by the extent to which they had institutional recognition from the public authorities.

The organisations taking part were subject to the arbitration of the mayor. This process of arbitration took as it starting point a list of associations, unions and public institutions having an interest in the project for the redesign of the public open space – the TNP, the post office, the local police force, the town hall, the employment bureau, housing organisations, and local retailers. All the neighbourhood committees were invited to take part and were represented by several of their members. These representatives were recognised as users of the Place Lazare-Goujon. Other more specific users were also linked to the project through the representation of groups for those with mobility impairment and for pedestrians. This list of organisations was drawn up by the organisers of the resources group. It was prepared in an 'intuitive manner' (technical officer, working group meeting 5 February 2003), in that there was no research done to establish exactly who might represent users of the square, and the organisations were invited to send a representative (mediator, working group meeting 5 February 2003).

The political manoeuvring was quite explicit in some of the working meetings: some associations were effectively excluded from the group.[11] Such manoeuvring was in part responsible for the time needed to draw up the list of legitimate and effective members of the resources group – over four months. The choices made influenced who was represented on the group and the representation was criticised at group meetings. Young people, for instance, were poorly represented. The resources group met seven times between May 2002 and March 2003 at Villeurbanne town hall, in the council chamber, from 6:00 p.m. to 8 p.m. The timing of these meetings was largely due to the elected representatives' timetable, and particularly that of the mayor, who might have to attend two meetings in one evening.

In terms of the formal procedures, minutes were kept of the first six of these meetings. The location of the meetings was the subject of discussion between the organisers. The council chamber in which most of the meetings were held has a high ceiling and a table equipped with microphones. It possessed a certain formality, which reflects the importance of the decisions that are usually taken there. But the chamber appeared unsuited to the probable form of these debates. Indeed, the layout of the space appeared to have a significant effect on the discussion that took place,

---

[11] At one meeting, an elected representative said, 'our relations with the MJC were poor. It's all right for potter's workshops ... but there are other organisations around' (elected representative, working meeting 13 February 2002). The MJC was not represented on the resources group.

inducing moderation in some participants and reducing the more timid to silence.[12] Meetings held in more suitable and less formal surroundings appeared to make it easier for people to have their say.[13] Meetings of the resources group did not all have the same character: the subject matter dealt with and the objectives promoted developed as the place Lazare-Goujon project itself developed. These variations made it possible to construct a typology of these meetings: an initial contact meeting, four briefing meetings, an evaluation meeting, and a meeting with the designers.

The initial contact meeting. This initial meeting on 16 May 2002 was the starting point for the framework and was intended to present the objectives and working methods of the resources group. Participants were placed in a precise order. In the council chamber, the tables were arranged in a rectangle. The first row of chairs made up a first circle, which widened into a secondary semi-circle on the west and south sides of the table. The first circle, together with those in the second semi-circle, had microphones.

The meeting had three phases: a presentation by the politicians of both the public open space and participation projects; a technical presentation of the participation project; and a round-table discussion.

The briefing meetings. Four meetings had been arranged to draw up the brief for the public open space project. They took place between May and October 2002, between 6:00 p.m. and 8:00 p.m., in the council chamber. The seating of participants was not explicitly made an object of proceedings and was left open. At some meetings, name cards with the names of the invited participants were available on the table by the entrance and everyone was invited to take their own. Sometimes, the name cards were already put on the table in the chamber, but in that case the participants took it upon themselves to choose whom they wanted to sit next to by moving their card.

By observation, in spite of the freedom which the participants had, seating arrangements varied little from one meeting to the next; elected representatives occupied the west side of the table, the mediator the east, the users' groups representatives the north and south sides, and the researchers set back from the main rectangle on one of the rows of seats without microphones. The elected representatives placed themselves at the head of the table and chaired the meetings. Some participants were excluded from the discussions, by virtue of placing themselves, like the researchers, at tables without microphones.

---

[12] Opinion was divided among residents. 'The place is very formal ... and that's all right ... there, the mayor has to reply ... the town hall deals with the heart of the town ... there's a right size for everything ... somewhere else would have been better ... the social centre, for example ...'. Another participant made the same point in an aside (resources group meeting, 5 February 2003).

[13] This is conjectural and cannot be scientifically verified, given the number of meetings observed. It is simply an observation by the researchers on the significant differences of ambiance – where people sat and the way that objects were divided among them – between the meetings in the council chamber and the meetings in adjacent rooms which were smaller and less self-consciously formal. It is also true that not all the meetings had the same objectives, and therefore were inevitably different in character.

Drawing
showing the
square

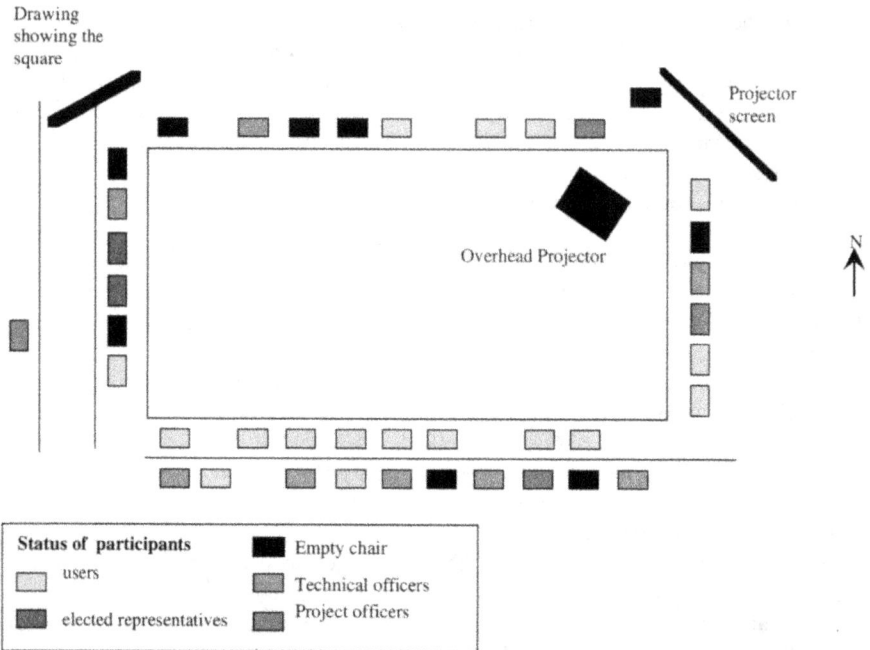

Figure 11.3 Plan of the Council Chamber, Meeting 16 May 2002

Formally, there were two parts to the meetings: a presentation of the plans for the public space – in particular the place Lazare-Goujon – and a discussion. The first presentation emphasised the square's history and character. The approach and the presentation itself were the work of students on the professional postgraduate course (*DESS Urbanisme, Aménagement et Gestion de la ville*) at the Institut d'Urbanisme. The second presentation 'took the social science approach that was integral to the programming of the brief for place Lazare-Goujon' and consisted of explaining the form and structure of the programme for the open space. The third presentation was devoted to a sociological study of the uses to which the square was put. Private consultants had undertaken this study. Finally, the meeting was the occasion for presenting the brief (or programme) for the project for the open space to those design teams invited to tender for the design.

The monitoring meeting and the meeting of the designers. The purpose of the monitoring meeting, which took place on 5 February 2003 from 6:30 to 8:00 p.m., was to evaluate the work of the resources group in the four previous meetings.

The meeting of the design teams was held on 25 March 2003 in the council chamber from 7:30 to 10:00 p.m. Its purpose was to allow the teams to present their schemes to the resources group. Elected representatives were present in force and had much to say; indeed, to a large extent, they dominated the proceedings. The schemes were not to be submitted formally until ten days later, on 4 April 2003, and at the meeting, the design teams had to remind their audience on several occasions that proposals were not yet finalised.

The Public Exhibition

A public exhibition was held in the entrance hall of the town hall at Villeurbanne from 20 May to 30 June 2003, between the two rounds of the selection process. The four schemes were presented alongside a brief history of the square. Visitors to the exhibition were invited to complete a questionnaire so that residents' and users' preferences in relation to the development of the square and the objects within it could be gauged. The team of sociologists who had undertaken the sociological survey were available at the exhibition two days a week. The team also analysed the results of the visitor questionnaire on behalf of Villeurbanne council.

The study of the arrangements for public involvement aimed to establish what the real effects of participation had been both on the network of actors involved in the project and, as a result, on their ability to take decisions. It was a question of taking account of the technical and spatial effects of participation: in what ways did participation affect the form of the project and its implementation?

Participation and the Possibilities of New Forms of Information Exchange

The way in which actors are mobilised is important, and in the Place Lazare-Goujon project this mobilisation depended on a number of more or less explicit rules. The composition of the resources group was the subject of long discussions among those promoting the participation programme. The representation of councillors, technical officers and residents needed to proportional. The days and times of meetings had a strategic importance, in that the choice could result in some elected representatives and members of the public being excluded, for instance, should the meetings take place in working hours.

Besides the composition of the group and the planning of the process, the arrangements for participation allowed rules to be devised – about who spoke and when – on the distribution of objects to members of the group – as well as for the stakes for which the actors were playing, to take the analogy of gaming[14] (Goffman 1973). These rules were explicitly designed to focus discussion on the car park project and the consequent improvement of the square. They were devised for the most part by elected representatives and technical officers before the arrangements for participation had been determined. The programme for participation simply consisted of demonstrating what the rules were. At the inaugural meeting, the elected representatives' introductory remarks were geared to establishing the rules while setting out the principles on which the development was based – and by so doing, identifying what was, and what was not, negotiable. Councillors expressed themselves clearly on the technical and spatial parameters of the project that could be the subject of participation and by the same token, they defined the elements that were to be

---

[14] In traditional games the 'objects' correspond to cards or to pieces; in this case, by 'objects' we mean plans, project reports, etc.

excluded from the process of participation.[15] Finally, the elected representatives and the technical officers defined the themes of future meetings.

Figure 11.4 Exhibition Poster

By establishing rules in this way, elected representatives and technical officers fixed the objectives of the resources group: '...to feed the brief ...'; '...to inform the brief ...'; '...to make as successful a public open space as possible ...' (resources

---

[15] This observation seems to reflect Jeudy's remark that '... what is not negotiable is, paradoxically, the starting point for the act of negotiation' (1996, p.14).

group meeting, 16 May 2002). These rules also seemed to determine the spatial arrangement

of participants in the council chamber. From one meeting to the next, participants in the resources group took up the same positions, starting with the councillors who, naturally, chaired these meetings.

The existence of these rules, in particular those that were implicit, became apparent when participants tried to behave differently. For example, residents' representatives wanted to ask for a model to be produced and an additional meeting at which the model could be discussed: '…the square is increasing in size from 5000m² to 10,000m² … the trees do not work … they should be changed … we need a meeting with a model to understand what is happening … I really feel the lack of that … so many new things … don't we need an additional meeting?' The technical officer present replied: '…that really isn't possible … we can't work like that … we really couldn't agree to that … we must let the professionals get on with it … there'll be a chance for you to react later … it isn't possible for us all to work together' (resources group meeting 4 July 2002). The residents tried to get round the technical officers by approaching the organisers and notably the elected representatives direct, but they got the same response: 'there can't be a model' (technical officer, resources group meeting 5 February 2002). In the same line of argument, a representative of one of the groups asked whether a public meeting for residents could be held: '…there's been a request …' he argued, '…this is the second time … our group met a fortnight ago …'. He went on: '…representatives here are not the majority … they want to have their say … perhaps we should really have a democracy …' (resources group meeting, 3 June 2002). The councillor present accepted the principle of such a meeting, but the meeting never took place because the majority of elected representatives and technical officers objected to it in other places.

Discussions about the rules brought to light dissension between those directly involved in the project and those affected by the development. Contradictions appeared which led to actors involved justifying their positions or a taking a critical stance: '…at Villeurbanne … the purpose of participation … of the resources group … is co-production … it's about reaching a consensus … the resources group … is too small … there will be only five meetings … it's too short … to explore the contradictions … and then find a consensus … the rules of engagement are too strict because of the constraints of time … and means … it's a small project … the objective of participation is political [...] the group's purpose is to gain experience in participation before the project begins and to establish links with the neighbourhood councils [...] the project is not in fact the object of participation … it's a risk-free project … as for participation … there's no ideal solution … the mayor is looking for opinions … is wanting to show that councillors and technical officers are listening … there's no demand' (technical officer, resources group meeting, 5 February 2003). To one resident who suggested that '…we need to participate in the decision making even if we are not actually taking the decision …', the officer replied: '…the last time we reached agreement … you gave your opinion …' (resources group meeting, 5 February 2003).

But these dissensions and contradictions did not seem to have a major impact on the satisfaction levels of members of the resources group. These members, particularly those who were neither elected representatives nor officers, did not object to either the way that participation had been set up or to its results. Indeed, on the contrary, they declared themselves satisfied by the programme for the project and gave their support to the mediator. The way in which the participation had been structured had, in this light, worked perfectly, keeping disagreement within limits and outside the resources group: '...the work of [the sociologists' team] ... was good ...'; '...taken together ... comments were taken on board ... we have to thank the planning agency for having worked in that way ...' (users and residents, resources group meeting, 1 October 2002). Or again: '...I really liked the mayor's contributions ... the mediator was good ...'; 100 per cent good ... wonderful ... the technical input really put us right ... made things clear ... gave us ideas ...'; '...the participation was innovative ...' (users and residents, resources group meeting, 5 February 2003). The mediator's role was particularly appreciated. Recognition for his input seemed to relate to the fact that he was seen as guaranteeing the discussions: '...I really want to thank the mediator ... everyone was able to express themselves freely ... everyone respected everyone else ...' (participant, resources group meeting, 1 October 2002) '...very, very good ...'; '...a counterweight in the gathering ...'; '...he picks up ideas ...'; '...the mediation allowed everyone to be heard ...' (users and residents, resources group meeting 5 February 2003).

This general satisfaction did not prevent participants realising that they were participating to very little effect: '...we're participating in the project for the surface of the square ... but nothing can be shifted ... just deciding on the colour of litter bins ... is of no interest ... we were not involved in the underground services or the car park and in the financial arrangements ... we are participating in nothing [...] we've nothing to play for ... things had been fixed before [the participation process] ... the car park ... but there were minor problems left ... I thought we'd been rather caught out by the participation ... participation didn't involve very much at all [...] I thought that we could enliven the square by bringing new elements ... a paper-stall for example ... but the mayor was flatly against it ... can you really talk about participation if the mayor is giving his opinion? [...] it's a lot of effort for a small square ... consensus was to be expected because everything had been fixed in advance ...' (users and residents, resources group meeting, 5 February 2003).

The Impact of Participation on Technical Decisions

The public participation process for the development of Place Lazare-Goujon were built on a consensus on the basic principle of 'reconstructing the square in its original form' that was defined by elected representatives, officers and the historic buildings architect. All the participants approved this principle, and no one thought eliminating cars would modify the appearance or the use of the square. Indeed, the programme proposed by elected representatives and officers seemed very close to the expectations of residents – at least those who expressed an opinion.

Elected representatives explained this closeness of view in this way: 'Place Lazare-Goujon and the tower blocks are in a ZPPAUP ... this brings with it a number of constraints ... the regulations for this zone date from the 1980s and

90s[16] ... nothing can be changed ...' (mayor, resources group meeting, 16 May 2002); '...[the square] can only be changed very slightly and must be welcoming and central ...' (councillor, resources group meeting, 3 June 2002); '...I don't want to see it enlarged ...' (councillor, resources group meeting, 4 July 2002).

Technical officers took up the theme of the constraints on change. For the director of technical services, '...the square is seen as working well by most of its users ... that's not common ... we must show a certain humility ... take what exists as our starting point ... we need to maintain the place ... the square is very harmonious ...' (resources group meeting, 4 July 2002). For the historic buildings representative '...we need to rediscover the simplicity [of the original design] ... it mustn't be turned into a museum piece ... or be modern for the sake of being modern ... we need to keep some strong design elements ... the overall symmetry ... the proportions of the pools in relation to the square ... the pergolas and the corner treatments ... this composition is based on a number of essential elements ...' (resource group meeting, 1 January 2002).

Residents who expressed an opinion confirmed the thrust of this discussion: '...people I've met don't see the need for change ... it's a matter of improving ... not changing ... we've debated the issue a fortnight ago ... the car park's going to improve a lot of things ... there's very little need to change the square ... it's the correct shape ... in the same way one might talk about something having a correct flavour ... let's keep it that way ...' (users and residents, resources group meeting, 3 June 2002).

The three elements of the participation process – the public meeting, the public exhibition, and the resources group meetings – allowed participants to find out about the different elements of the square. The numerous questions asked did not give rise to elements that could be directly incorporated into the design or into a restatement of the constituent elements of the square. But they did amount to a body of material, which was sufficiently rich to give an insight into the proposals, and objects that participants felt concerned about: '...will there be motor-cycle parking under surveillance? ... what about the problem of water in the car park?...' This question was prompted by the technical solutions, particularly pumping, found to deal with the height of the water table relative to the depth of the car park. '...The car park has too few spaces ... how many spaces are represented by the 48 per cent reserved for companies?...' (users and residents, resources group meeting, 28 November 2002); '...you've talked about reducing the size of the pools ... how do you intend to do that?...' (users and residents, resources group meeting, 19 June 2002); '...the police station is going to be located in the France Telecoms office ... we need a way out ... it would be good to have an access road (police representative, resources group meeting, 4 July 2002); ... are the pergolas going to be kept?...' (user, resources group meeting, 1 October 2002); '...where is the war memorial going to be put?...'; '...what form of lighting is proposed? ... what will the ground treatment be ...?' (users and residents, resources group meeting 25 March 2003) and so forth.

Beyond information-seeking questions, there was little discussion on the physical arrangement of the square: the benches, the pools, the porches, the fountains, the bust

---

[16] In fact, the ZPPAUP was declared in 1993.

of Lazare Goujon, the trees and the ground treatment.[17] An exception was the war memorial, the moving and siting of which was the subject of comment (resources group meetings, 16 May 2002, 3 June 2002, 19 June 2002).

At the very first meeting, a representative of a veterans' group touched on the question of the war memorial: '...I want the memorial to remain in place ...' (resources group meeting, 16 May 2002). He followed up the theme at the second meeting: '...about the war memorial ... perhaps it could be moved ... moving it could get over the problem we have at commemorations ... the flag-bearers and the public can't stand aside ... so the public can't hear the flag-bearers ...'. The project leader said: '...be careful ... we'd need the approval of the families if the memorial is going to be moved ... it contains funerary caskets ...' (resources group meeting, 3 June 2002). When the competition schemes were presented, this question re-emerged after the presentation by each of the architects: '...where is the war memorial...?' (resources group meeting, 25 February 2003).

Users and residents also expressed their expectations on the uses to which the square would be put. Elected representatives and officers then put their point of view. In fact, there was no real discussion and the interactions[18] that engaged actors were very short, amounting to no more than one or two exchanges. Residents frequently opened these exchanges by introducing the subject. Elected representatives took charge of the closing sequence, by having the last word. Thus in the course of the meeting, one participant said: '...the car park would be better placed there ...' and the mayor replied: '...anyone can say where the car park ought to go ... but now ... the car park is fixed ... there are things we can discuss and things which are already fixed.' (public meeting, 28 November 2002).

Another question of the activities, the use of the entrance to the square during wedding ceremonies, gave rise to debate over several meetings (resources group meetings, 16 May 2002, 4 July 2002, 1 October 2002). This engaged elected representatives, officers and residents in discussions about the best spatial arrangement for ceremonies, given current difficulties. At the meeting on 1 October 2002, for example, councillors reckoned that as far as access to the town hall for weddings was concerned, '...we need to be very careful ... we all know the difficulties with weddings ...'. Comments by the residents made the difficulties with the means of access clear: '...the north entrance ought to be brought back into use ...' said some, while others said, '...we don't agree about the north entrance ... the north entrance lacks the character of the south entrance ...' – and it was the issue of 'character' that was at stake in the disagreement. In order to bring the south entrance into use, residents made a technical proposal: '...we suggest an automatic barrier.[19] The mediator intervened to close the discussion by saying '...this is a detail of the

---

[17] At the meeting with the designers, some of these elements were touched on, but mainly by the elected representatives.

[18] 'Interactions' are taken to be a 'communicative entity' that presents an evident internal continuity, both of the group and of the matter treated, and that marks a break with what precedes and what follows it (after Kerbrat-Orecchiani 1996, p.36)

[19] The proposal was for a rising bollard accessible only for wedding processions on the roadway between the south front of the town hall and the north side of the square, requiring a code or card to allow access.

scheme ... it's up to the designers to propose a solution ...' (resources group meeting 1 October 2002).

In fact, discussion about the use of the square was limited. As soon as this type of topic was broached, councillors moved to defer the discussion. For example, when the project leader for the square talked about use at the last meeting before the brief was finalised, councillors present told him that the discussion would come later: '...I think the debate will happen later ... we need to see how the square will shape up without the two routes through ... we need to see how the central uses change [...] how will the central uses be transformed by the transformation of the square? [...] we really need to think about that ...' (elected representatives, resources group meeting, 4 July 2002). This stance on the part of councillors seemed – to the external observer – all the stranger for the fact that the development of the square would involve changing nothing of the existing activities would, rather, result in an improvement to the way in which the square was being used already, as typified by the possible relocation of the war memorial. The position of councillors, in as much as they were the ones who were the bearers of the participation process, was that the purpose of participation was not to anticipate the future uses of place Lazare-Goujon. In other words, if participation were to change anything with regard to the layout and technical detail, this should not entail a consideration of the adequacy of the development for activities. Participation was to be about the spatial arrangement and the technical realisation of the development. Thus the question of the practice in respect of weddings could only be resolved spatially and technically by proposals from those in charge of the work, and they alone – not the residents – could propose technical and spatial solutions to the problem.

Conclusion

Only limited conclusions can be drawn from this observation of the participation process in Villeurbanne – limited because the whole process was devised, and its implementation and effectiveness was tightly linked to Villeurbanne and the open space to be improved. But it is also true that this participation constituted the first step in the in the renewal of democratic process based on the principle of 'participative democracy'. To that extent, the particularities of the process can shed light on the general application of participative democracy.[20] The experiment in participation in the place Lazare-Goujon project was in the end limited to a strictly controlled exercise: the square was a major element in the town, but in view of the constraints of the site, its development offered little in the way of participation, because in order to move swiftly, nothing was changed. The institutional setting as well as the procedures applied depended entirely on an existing politico-administrative structure and the usual rules that applied to meetings. The selection of participants

[20] See also the Charter for participation in Greater Lyon, and in particular, clause 3 'to engage in a dynamic of participation'. Indeed, the Charter's objective – and as a consequence political action – was 'participation', that is to say 'a dialogue with the population on any particular subject, in order the better to take into account opinion and so to establish the general interest' (Grand Lyon 2003, p.8).

depended on criteria that remained vague. In effect, there were no new forms of organisation or procedures.

In spite of these limitations, observations made do not allow accusations of bad faith or hypocrisy to be made on the part of those who organised the participation process. On the contrary, the good faith of those engaged in the process was incontestable. Indeed everyone involved was agreed on one point: that participation should be successful and that the process should be exemplary. The actors in the process behaved rather like those in charge of clearing minefields when faced with a mine, which in the words of some of the elected representatives and officers '...risked blowing up in our faces...'. In this initial experiment of what was intended to become general practice in participative democracy,[21] the actors involved in the project for place Lazare-Goujon all followed the same strategy. In order to embed the participation process, they all supported the idea of modest participation that would not give rise controversy or violent confrontation between participants in order to underwrite the credibility of the first great experiment.

This last observation opens up other lines of thought. We can ask whether this strategy is not the effect of a belief in the possibility of taking action, of doing, of 'making' without taking responsibility for the consequences. Such a belief makes it possible to act without power, that is to say, without the risk of having to choose a course of action, in the understanding that it is possible to establish a world devoid of controversy, confrontation and conflict. The best of all worlds would be one in which consensus reigned, a world without need of making choices because a single solution would emerge of its own accord, a solution that participation, amongst other things, would help to reveal.

The limitations of the place Lazare-Goujon experiment lay in the way in which the workings of democratic and political institutions were represented. Democracy would end conflict in the endless search for the 'right way', which would be beyond conflict because it would involve no controversy. In other words, this mode of representative democracy is a way of getting round another representation that sees democracy as a means of articulating conflict, but by peaceful means. In the first case, the existence of consensus presumes the conditions under which a fair and legitimate design would be recognised perforce by society as a whole. In the second case, the existence of conflict is no more than the recognition of the diversity of the world, of the choices open to everyone or, in the sense offered by Castoriadis (1975) of the possible production of otherness, or again the production of the new.

References

Azéma, J.-L. (2001) L'expérience lyonnaise, la naissance d'une organisation. In *User, observer, programmer et fabriquer l'espace public*, sous la direction de J.Y. Toussaint et M. Zimmermann (eds) Coll. des Sciences Appliquées de l'INSA de Lyon, Presses Polytechniques et Universitaires Romandes, Lausanne, pp.185-198.

---

[21]Since the setting up of the resources group for the place Lazare-Goujon project, two other groups have been set up in Villeurbanne.

Blanc, M. (1995) Politique de la ville et démocratie locale. La participation : une transaction le plus souvent différée, *Les Annales de la Recherche Urbaine* n°68-69, pp.98-106.

Blondiaux, L. (2001) La délibération, norme de l'action publique contemporaine?, *Projet*, n°263, pp.81-90.

Callon, M., Lascoumes, P. and Barthe, Y. (2001) *Agir dans un monde incertain. Essai sur la démocratie technique*, coll. *La couleur des idées*, Paris, Seuil.

Castoriadis, C. (e1975) *L'institution imaginaire de la société*, coll. Esprit, Paris, Seuil.

Goffman, E. (1974) *Les rites d'interaction*, coll. Le sens du commun, Les éditions. de Minuit, Paris.

Grafmeyer, Y. (1994) *Sociologie urbaine*, Paris, Nathan.

Jeudy, H.-P. (1996) Prologue, *Autrement*, series *Mutations n°163*, pp.12-17.

Kerbrat-Orecchiani, C. (1996) *La conversation*, coll. Mémo, Paris, Seuil.

Meuret, B. (1982) *Le socialisme municipal. Villeurbanne 1880-1982*, Lyon, Presse Universitaire de Lyon.

Midler, C. (e1998) *L'auto qui n'existait pas. Management de projet et transformation de l'entreprise*, Paris, Dunod.

Mollet, A. (ed.) (1981) *Quand les habitants prennent la parole*, Bilan thématique Plan Construction, Paris.

Neveu, C. (1999) Espace public et engagement politique. Enjeux et logiques de la citoyenneté locale, coll. Logiques Politiques, Harmattan.

Paoletti, M. (1999) La démocratie locale. Spécificité et alignement. In CRAPS-CURAPP (ed.) *La démocratie locale. Représentation, participation, espace public*, Paris, Presses Universitaires de France, pp.219-236.

Ragon, M. (1977) L'Architecte, le Prince et la Démocratie. Vers une démocratisation de l'Architecture Paris, Albin Michel.

Toussaint, J.-Y. (2003a) Le projet d'aménagement : mobilisation d'acteurs et institution d'un collectif d'énonciation, *Recherches Transversales. Cahier de Recherches du Centre des Humanités* n°7, dossier Le management de projet, de quoi parle-t-on?, pp.59-77.

Toussaint, J.-Y. (2003b) *Projets et usages urbains. Fabriquer et utiliser les dispositifs techniques et spatiaux de l'urbain*, rapport de HDR, Y. Grafmeyer, ed, mimeo.

Toussaint, Jean-Yves, Vareilles, S. and Zimmermann, M. (2002) *L'aménagement des espaces publics comme mise en œuvre de la démocratie. L'expérience lyonnaise de l'aménagement des espaces publics*, research report for the Ministère de l'Equipement, des Transports et du Logement, mimeo, Institut National des Sciences Appliquées, Lyon.

Toussaint, J.-Y. and Zimmermann, M. (2001) De quelques difficultés à prendre en compte les usages dans la conception de produits. Le cas des dispositifs techniques et spatiaux de l'urbain. In Perrin Jacques (ed.) *Conception, des pratiques et des méthodes*, coll. des Sciences Appliquées, Lausanne, Presses Polytechniques Universitaires Romandes, pp.215-238.

Chapter 12

Local Democracy Under Challenge: The Work of the Agora Association in Vaulx-en-Velin, France

Didier Chabanet
Translation by John Tittensor

Since the mid-1970s 'the city has replaced the workplace as the main theatre of social conflict' (Chaline 1997, p.8). As many analyses have already made clear, this shift is a significant one, giving expression notably to the combined effects of mounting social inequality – more and more obvious in the geography of our cities – and a growing rejection of the traditional forms of commitment, be they political or trade-union based. Disadvantaged neighbourhoods are a core part of these changes: often of immigrant stock, the residents generally see themselves not as victims of exploitation but as social rejects. In this respect, then, the nature of the immigration question has changed considerably and in doing so has cut free from the world of work and industrial conflict. The dominant issues now are the right to a full life, to success, and to individual participation, and as such are increasingly tied to the urban question as a whole. It is this change that has led the authorities in France, and first and foremost the state, to implement a series of initiatives making up what is called 'la politique de ville' ('community development and urban regeneration policy').[1]

The first measures were largely housing rehabilitation operations,[2] rapidly extended in the early 1980s into a vast programme of neighbourhood social development.[3] There followed a host of initiatives,[4] with the 'politique de ville' tending to evolve into a complex system of benefits involving many different

---

[1] In December 1990, the creation of a Ministry of Urban Affairs provided visibility for measures aimed at state-classified 'disadvantaged neighbourhoods', most of which are made up of social housing.
[2] Via the 1977 'Housing and Community Life' agreements.
[3] This programme gave broad expression to the preparatory work of the Bonnemaison, Schwartz and Dubedout commission, charged respectively with making suggestions for crime prevention, integration of the young into the workforce and improvements to social housing.
[4] Notable among the main measures were the creation in 1988 of the Interministerial Delegation for Urban Affairs and the National Council for Urban Affairs, the 1990 'Besson law' on housing for the poor, the 1993 procedure for state-city planning contracts and, more recently, the Urban New Start Pact of 1996.

ministries. The state, however, was concerned not simply in distributing resources, but also in bringing about a significant shift in France's politico-administrative ethos. According to Jacques Donzelot and Philippe Estèbe (1994), public action had evolved from a national programme to local project, from procedure to contract, from administrative management to motivation of local actors, and from intervention area to neighbourhood mobilisation.

The scope and political ambition of this system doubtless account in large part for the ambiguity of aims so often highlighted by evaluators (Bélorgey 1993). In particular, community development and urban regeneration policy seemed to oscillate between two agendas, which, without being mutually exclusive, were not equally effective. The first had to do with improving the living environment, housing, town planning works, and infrastructures; while the second, bent on more overt political correctness (Fontaine 2003), made extensive use of notions like 'democracy', 'citizenship' and 'participation', especially in its dealings with community association actors in disadvantaged neighbourhoods. The divergence between the two registers became almost a call to order: 'While action conducive to urban development and rehabilitation and improvement of the urban habitat must be pursued and enriched so as to respond better to existing needs, it cannot be the centre of gravity, the main emphasis of the 'politique de la ville'. The core aim must be integration of residents via economic and social development of disadvantaged neighbourhoods and areas, and the creation of a citizen identity, with young people as top priority' (Sardais 1990, p.93).

It is this aim that I would like to analyse critically, drawing on the opinions of those who are first and foremost concerned as the target of these measures. It is no longer a source of real surprise that the relationship between the population as a whole and a given policy is marked by considerable ignorance and/or indifference (Marie et al. 2002). Yet we may well wonder why community association actors, thoroughly up to date on current institutional mechanisms, slate the 'politique de la ville' so vigorously – while remaining ready, when the occasion arises, to get the most out of the budgets and other assistance it makes available. My observations are based directly on research carried out on the Mas du Taureau housing estate in Vaulx-en-Velin, near Lyon, among young members of Agora, the neighbourhood's main community association.

Since Agora's beginnings, its role has evolved considerably. It was initially established a few days after the riots of October 1990 as the 'Thomas Claudio Committee', from the name of a young resident who died in a clash with police. The committee's aim was to monitor the affair in the courts; the actual association was officially founded in November 1991, the intention at the time being broader action via denunciation of the authorities' chronic inability to meet the needs of residents – especially the young. Agora gradually became the main vehicle for the expression of youth discontent and the venue for development of young people's own initiatives. Its influence, then, is to be measured less in terms of membership numbers – a factor with no very exact meaning in this case[5] – rather than of the support it receives from

---

[5] As an indication, Association meetings were rarely attended by more than twenty people.

the local population: in Vaulx-en-Velin it is now a crucial actor on the social and political scene.

I shall begin by looking at the way immigrant integration fits with global changes in a society marked by the ongoing breakdown of its industrial framework (*1. From the social question to urban problems*). Next I shall show that the new problems, linked to acute forms of urban segregation, can become worse and sometimes, as in the case of Vaulx-en-Velin, lead to rioting (*2. Vaulx-en-Velin fragments: the October 1990 riots*). Working from this example, I shall analyse not only the grudges vented, but also the political demands formulated by some of the young. This will highlight the existence of two competing forms of legitimacy. The first, embodied by the young people of Agora, is based on resident perception of second-class citizenship as a lever for action and a criterion for justification of claims – basically regarding participation and autonomy – within a clearly defined living space (*3. Agora members and the aspiration to autonomy*). The other, defended by the municipality of Vaulx-en-Velin, is founded on the basis of universal suffrage, as sanctioned by the population as a whole, at municipal level. These are the two rationales that clash locally (*4. Creation of conflict on a municipal scale*). This conflict found temporary resolution at the time of the 1995 municipal elections, when Agora, backed notably by other local associations, instigated the creation of a broad-based ticket whose two main demands were resident involvement and improvement of living conditions in sensitive neighbourhoods (*5. The municipal election campaign of June 1995*). More recently, Agora's agenda has taken on a fresh slant marked by rapprochement with religious associations – in particular the Union of Young Muslims – as part of the MIB: (Mouvement de l'Immigration et des Banlieues / Immigration and Neighbourhood Movement) (*6. The formation of MIB*).

From the 'Social Question' to Urban Problems

From the late 19th century to the beginning of the 1970s, immigration was largely an industrial issue, the influx of foreigners being primarily intended to meet economic needs and remedy a chronic national shortage of manpower. The image of the immigrant as unskilled industrial worker was a widespread one. A major feature was the length of stay: 'It was thought that migrant workers, after several years of saving money and in response to market forces in their country of origin, would spontaneously return home, leaving room for other migrants. This return was seen as the natural culmination of migratory movements' (Schnapper 1986). This idea was current both in the host society, among migrants themselves, and often in the society of their country of origin. The immigrant was universally perceived as 'basically a provider of work – and a temporary one, a worker in transit. In the final analysis an immigrant's raison d'être is strictly provisional and dependent on his doing what is expected of him: he and his raison d'être exist through work, for work and in work; because he is needed, for as long as he is needed, for that which makes him needed, and where he is needed' (Sayad 1992). Thus reduced to its strictly economic dimension, immigration as such does not figure in political debate.

Today, however, the visible character of immigration has changed and it is only too clear that employment immigration has been replaced by residence immigration. Paradoxically it was the official halting of immigrant entries in 1974 that speeded up this change: 'The closing of frontiers ... radically modified immigrants' planning, reducing the incentive to return, increasing the length of stay and leading many – probably the majority – to envisage remaining permanently' (Tapinos 1988). From that date onwards the likelihood of a return home started to fade.

At the same time that integration of migrant groups began to stabilise, it changed character. Applied by Sayad (1977) to successive waves of Algerian immigration, analysis of these changes points up major sociological constants. Initially the migrant retains strong ties to his community – most often a rural one – and his country of origin. He is young and unmarried, and the aim of his being in France is to provide the family circle at home with the resources it needs. As a rule, this kind of immigration is brief. Under the influence of the community of origin, the migrant is regularly replaced by other members. In a second phase, migrants break free of this influence as they adopt the dominant French lifestyles and images. As it develops, the migratory process gradually escapes from the control of the base community: becoming older and staying longer in France, migrants are no longer community representatives and their links with the homeland become more fragile. The third phase is the extension of this distancing process, with immigration tending to become permanent and migrants arriving much more often with their families. The diachronic schema adopted by Abdelmalek Sayad stresses the shift from employment immigration – temporary and typified by the image of the unmarried manual worker – to a definitive, residential immigration highlighted by the presence of the young of the second and third generations.

The history of Vaulx-en-Velin is an eloquent illustration of this change. Its first phase is that of the march towards industrialisation that shapes most of the social relationships of foreigners and their integration. This is followed by a revelation of the rise of problems emerging from the processes of impoverishment and stigmatisation.

Until the mid-19th century, Vaulx-en-Velin was a poor market-garden village specialising in the cultivation of cardoons. Industrialisation began to develop with the construction in 1892-98 of the Jonage Canal, intended to provide hydroelectricity for the Lyon economy. Parts of the municipality became residential areas for manual workers and Vaulx-en-Velin gradually developed with the opening of more and more factories. The main driving force behind this growth was the Gillet group, which reacted to the decline of the traditional silk industry by setting up a clearing house for artificial textiles in 1911. Soon a host of factories in Vaulx-en-Velin and its environs were manufacturing the new fibres and it was this adoption of technical innovation that gave rise to what would become an international empire: the Rhône-Poulenc company.

Dependent as it was on steadily increasing manpower, the industrialisation of Vaulx-en-Velin is indissociable from the arrival of foreign workers. The first wave came largely from Poland and from Silesia in particular. The Armenian genocide brought dozens of families, just as Gillet was setting up. Large numbers of Hungarians and Russians also found work there. In the between-wars period Italian and Spanish immigrants, fleeing Fascism and civil war respectively, arrived. They were followed, after World War II, by Kabyes leaving Algeria for economic and/or political reasons, and then by large numbers of Portuguese, refugees from the Salazar dictatorship. In the 1970s, Rhône-Poulenc began a new search for workers, this time in North Africa. Coming just before the closing of the borders, this final wave mainly comprised immigrants from the Maghreb, sub-Saharan Africa and the Far East, with the rise of industrialisation and concomitant demographic expansion – foreign and local – becoming motors for accelerated urbanisation. As in France's 'red districts' social and political activity grew out of the development-related association of production and housing. As a logical consequence the management of Vaulx-en-Velin was regularly entrusted to members of the communist party. From the election of Victor Pinsard as mayor in 1929, to that of Maurice Charrier[6] in 1985, the municipality's political colour has remained unchanged.

This internally consistent social order is now a thing of the past: the working-class neighbourhoods, with their systems of social relationships structured by workshop and factory, have been swept away by change. A major factor in the destruction of this way of life was the creation of 'HLM' ('moderate-rental housing') estates, the first of which went up in Vaulx-en-Velin in 1953-59. People with very low incomes were little represented in this kind of housing. Upper management and the professions – theoretically not entitled to accommodation officially intended for the impoverished – were over-represented in relation to the population as a whole (7 per cent as against 6 per cent), while the example of middle management was even more flagrant (16 per cent as against 10 per cent). Even working-class HLM residents belonged to the most qualified and, relatively speaking, most comfortable fringe group. The arrival in the early 1960s of occupants for substantial quantities of new housing in the ZUPs (Priority Urbanisation Zones) did nothing to change this sociological distribution. The population still seemed as mixed as ever. The prospect of living in these areas was not then perceived as demeaning, many of the new arrivals being pleased to find themselves paying a reasonable rent for relatively comfortable accommodation.

Social balance in the ZUPs was short-lived, however. From the early 1970s, whole areas to the east of Lyon – and Vaulx-en-Velin in particular – gradually declined in status owing to the combined effects of a number of housing policy mechanisms. The first was the rise in private-sector rentals; the second was the regeneration of certain slum areas. These measures were accompanied by substantial rent rises and the loss of housing within the reach of the lowest income brackets. Steep increases in city-centre property values pushed the most disadvantaged groups towards unoccupied social housing; this was especially true of migrants, already more often in precarious situations.

---

[6] The present mayor.

Vaulx-en-Velin Fragments: the October 1990 Riots

This de facto segregation[7] had clear consequences within Vaulx-en-Velin itself, with the concentration in specific neighbourhoods, both of groups defined by their precariousness and of an accumulation of social handicaps influencing the perception of the community as a whole.   The difficulties of coexistence, mutual incomprehension, and the deterioration of the living environment led many people – when they had the chance – to leave for more congenial settings. Others left because their standard of living had risen and they wanted to buy a home. Principally, however, the housing estates were equated with social inferiority, even when this judgment had nothing to do with their actual state. For HLM residents, home ownership was the culmination of a dream and a measure of success.

Whatever their motivations and their plans, the wealthiest families disappeared from the ZUPs (Priority Urbanisation Zones) with many of them settling in Vaulx-en-Velin's residential areas. Social disparities are now a clear geographical marker in a municipality that seems more fragmented than ever before. With the exception of the central area, which includes most state and local authority offices and a major retail hub, housing zones show very marked contrasts. The Village, built around the old town, can be readily identified by its many small shops and single-family houses, and the indicators point to a fairly balanced social situation: unemployment is running at only 10.7 per cent; the number of employers and self-employed people is the highest in the municipality; foreigners make up only 15.3 per cent of the population; 46.3 per cent of residents are home owners; and only 23 per cent of those living in rental accommodation are in HLMs.

In Vaulx-en-Velin social difference is immediately perceptible in the urban landscape, where the setting changes character radically as you move from one area to another. One example is the way the irregular street pattern shaped over the years in the Village gives way to something much more geometrical. The ZUP approach has produced isolated, closed-off neighbourhoods characterised in the most obvious way by highly standardised architecture. The Mas du Taureau neighbourhood, like its neighbours, presents as a succession of tower blocks surrounded by parking lots and the remains of grassed areas. All the shops are concentrated around the central square and the entries to the towers are drab and often dirty and rundown. There is nothing to break the sheer monotony of shape and colour until you come to the straight main roads defining the outer limits of the neighbourhood.

Here too the visible divisions reflect a specific demographic distribution in total contrast with that of the Village. In Mas du Taureau unemployment is running at 25 per cent for the working population as a whole and 33.2 per cent for those under 25. The employment category that includes lower management, technicians, draughtsmen, sales representatives, primary and secondary teachers, social workers, nurses, civil servants, engineers, management, office staff, retailers and service-sector workers

---

[7] The meaning of 'segregation' varies widely according to author and discipline. For a global, critical presentation, see Jacques Brun and Catherine Rhein (Eds.), *La ségrégation dans la ville*. I use the term in Yves Grafmeyer's fairly classical sense: 'Simultaneously a social distancing and a physical separation' ('Regards sociologiques sur la ségrégation', in *La ségrégation dans la ville*, p.86).

represents only 2.9 per cent of the workforce. 98.7 per cent of all residents rent, and of them 96 per cent live in HLMs. The youth of the population is another striking feature: 40.4 per cent under 20 and 38.5 per cent are between age 20 and 39. The proportion of residents with an immigrant background – whether they are now French citizens or not – cannot be calculated exactly, but is undoubtedly high.[8] As a general indication, 31.6 per cent of residents are foreign nationals.

The existence of this largely young, socially disadvantaged and migrant population was driven home by the riots of October 1990. After the death of a young male resident of the Mas du Taureau neighbourhood in a clash with police, the media offered stunned France scenes of violence and looting that seemed to defy understanding. No amount of analysis alleviated this feeling of incomprehension. Nothing seemed to justify what was taking place: the sombre spectacle of young people determinedly destroying their own neighbourhood, and the increasing confrontations with the police.

This bafflement, sometimes spilling over into exasperation, was all the more pronounced in that from the early 1980s the neighbourhood had been the subject of a major enhancement project. The base concept of the project, however, was that of a social response to the difficulties experienced by residents of the most disadvantaged neighbourhoods: in other words, there was no question of official recognition of the existence of an ethnic or political community, or communities, with specific needs. In a Republic that is one, indivisible and secular, the newly visible problems posed by young people of migrant background in working-class neighbourhoods were considered the outcome of socio-economic inequality (Lapeyronnie 1993). Mas du Taureau, like other priority neighbourhoods, thus became the target of a host of social assistance measures and was even regarded by the authorities as a paradigmatic success. Most of the social housing units had been renovated or were on the point of renovation and observers of all political persuasions were unanimous in their praise of the efforts being made for local youth by a municipality devoting 60 per cent of its annual budget to sport and education. Only a week before the first clashes, a climbing wall had been inaugurated to all-round acclaim.

'Until that first Saturday in October, Vaulx-en-Velin had been an image of perfection. For local socialist MP, Jean-Jack Queyranne, for communist mayor, Maurice Charrier, for rightist Lyon Urban Community president, Michel Noir, the rehabilitation undertaken in this district of Lyon since 1985 was quite simply exemplary' (July 1990, p.2).

Agora Members and the Aspiration of Autonomy

The Mas du Taureau situation is perfect testimony to the benefits – but also to the uncertainties and, ultimately, the limitations – of the 'politique de la ville'. The basic question is why an area that had received so much official assistance became the focus for rioting. In their sheer unexpectedness, the riots are proof that the measures taken,

---

[8] Unlike many other countries, France does not question national origin in its census-taking.

whatever their scale, were on the whole inappropriate to the needs of many of the neighbourhood's young. The mistake had been to believe that more and better facilities would suffice to overcome the population's feeling of injustice, of the scorn endured, and the resultant need for 'recognition'.[9] In general terms the discontent and the demands find expression at two levels: a permanent hostility to the police – whose modes of intervention are often, for young people, indissociable from their experience of racism – emerges as the decisive factor, in that it largely structures the perceptions governing their contact with society at large; and then there are the multiple forms of social and economic exclusion, felt as stigmas and expressions of rejection. In such a situation 'it is illusory to seek to change the status of the space by improving its buildings if there has been no in-depth effort made to blot out the initial rejection by some symbolic indication that this neighbourhood, abandoned at birth, has at last been recognised and adopted by society as a whole' (Bonetti 1994, p.43).

Thus it is not so much money and facilities that the young members of Agora were demanding, as the right to assertion of their autonomy – to an agenda they saw as allowing them to decide on the meaning of what they do. 'In this sense, autonomy is indissociable from the experience of living in society. It is a notion covering individual and group behaviours, shaped by the heterogeneousness of their constituent principles, and by the activity of individuals who must create the meaning of their way of life within this very heterogeneousness' (Dubet 1994, p.15). For the young members of Agora these modes of understanding were very much identified with the standards embodied by today's communication and information systems, with an emphasis on 'self-fulfilment' (Schudson 1994).[10]

This mass individualism has two primary dimensions. One has to do with the notion of freedom, authenticity and self-realisation, and thus with a penchant for hedonism. The other, highly moral, calls for respect, dignity and tolerance in the context of the universality of the human being. These values are neither unequivocal nor homogeneous, yet they culminate in what a growing number of young people, at Mas du Taureau and elsewhere, have chosen as a Singular Being: one who asserts his own identity and his right to autonomy of both judgement and action. This change represents a radical shift in the relationship with the world. Any dominant cultural model tends to make individual experience a meaningful world in its own right, with the social actors seeing themselves as deploying enhanced awareness and control of

---

[9] As Charles Taylor emphasises, 'What is new about the modern era is not the need for recognition, but the possibility that this need might not be met. Recognition used to not be a problem: it was based on a social identity which itself was founded on social categories everyone took for granted.' (1994, p.56). Here we have the key to the difference between the 'sensitive neighbourhoods' and the 'red districts' of the past, which were sufficiently socially and politically structured to produce feelings of identity and recognition in which the individual and group dimensions were closely associated. Today, in sensitive neighbourhoods as in the rest of our global society, recognition is more and more perceived as something produced, or a victory on the part of the individual.

[10] Basically, adopting the ideas of Daniel Bell (1979), the author shows that a relatively homogeneous 'cultural sphere', increasingly internationalised and dissociated from the social structure, is tending to become a common norm of integration.

their lives, at the same time as the broadening of the frame of reference leads them to seek emancipation from current norms. Thus the relationship – in hierarchical terms – between the individual and the system becomes increasingly difficult and affects the traditional balance of social regulation structures as a whole. These manifestations of cultural individualism express an urge to lay down new – perhaps radically new – rules of commitment based on human, local considerations, together with a will for transparency and authentic participation. In both its reflexiveness (the concern with retaining control of one's commitment and having one's demands met) and its emotional content (the need to assert the authenticity of one's personality), this agenda leads to a (re)definition of democracy, now perceived as 'recognition of the right of individuals and groups to be the actors in their own history' (Touraine 1994, p.34). In disadvantaged neighbourhoods this aspiration is all the more intense for not being satisfied: in a society in which the rights of individuals are endlessly proclaimed and investigated, any frustration is felt with extreme acuteness.

In the light of this situation, the 'politique de la ville' contained the seeds of disenchantment and protest with intervention by the authorities often being seen – unlike performative discourse – as an obstacle to resident emancipation and participation. The harshness of this reaction was explained by the fact that apathy, discouragement, withdrawal and – even more seriously – rationales of self-destruction were assuming disquieting proportions. For Agora, then, it was not simply a matter of denouncing an external set of measures that would not provide satisfaction, but of resisting what is sometimes described as an 'enemy within' that weakens and even crushes residents' capacity to take the initiative.[11] Thus the 'politique de la ville' was broadly challenged in the name of the very principles of 'democracy', 'citizenship' and 'participation' it proclaimed. To take one example: its budgets allowed for the funding of community associations, but these associations were required in return to employ social workers, and thus as a rule lost their room for manoeuvre, finding themselves in a subordinate situation in relation to the municipal authorities. Conversely, the demand for real involvement seemed an implicit component of the creation of such associations:

> In essence the Agora concept is the appropriation of the living space via debate and the construction of ideas, then the realisation of these ideas at the initiative of the residents. The difficult context we live in shows how vital it is to encourage citizen and resident involvement in public life. Now more than ever we need concrete, effective measures for making residents of this neighbourhood alert to their responsibilities... Citizenship for us means being consulted on local projects that directly affect our everyday lives.[12]

Concerned not to settle for facile criticism and the attendant trap of mere posturing, the Mas du Taureau community association actors sought at all times to

---

[11] See François Dubet (1987): 'There are many ways of describing this messy situation, but all of them stress a shifting, contradictory character that is not readily perceptible because the actor himself seems lost in a weakened time frame, in wavering aspirations, in an indefinable ambivalence' (p.9).

[12] Agora document.

make real partnerships with those for whom they were taking action, involving them closely in project design and realisation. Repeated reaffirmation of this intention made it a theme that gave meaning to the overall set of measures, with a high level of involvement by the beneficiaries emerging as vital to their sense of satisfaction. This is why those in charge of the association decided against extending its activity to areas outside the Mas du Taureau neighbourhood: for the Agora president, regularly called on by residents of other estates in the ZUP to broaden the association's geographical reach, each person must take responsibility for himself in his place of residence and avoid all dependence on outside structures. In this respect the legitimacy of Agora's work is based on a principle running counter to that of political and administrative bodies, which generally operate 'in the name and on behalf of communities extending beyond the persons physically concerned' (Pharo 1990, p.390).

Creation of Conflict on a Municipal Scale

In making an imperative of actor autonomy, this approach led to a localisation of the issues. Two main considerations combined here, the first of which was a general loss of credibility by political ideologies and organisations, divided more or less on Left/Right lines. For young Agora members these distinctions were virtually meaningless, both camps being perceived as monolithic, hostile and composed of self-seeking individuals they held responsible for the deterioration of their living environment. Secondly, decentralisation had significantly modified the political balance, with mayors and local councillors taking on a more important role than members of Parliament. Ultimately this meant that the municipality was seen as the main culprit and was made the target of condemnation by the young. In this respect the failure of the 'Beur movement', made up of young people born in France of North African parents, was clear: the cooperation (or alliance) between local actors, trade unions and political groups, which in the early 1980s had led to nationwide networking and rallies culminating in the 1983 'March for Equality and against Racism', had run its course. France Plus and SOS Racisme, the two associations which, working differently and sometimes in competition had become the spearheads of this strategy, had lost all credibility at local level. At Agora, where national political commitments and divisions were seen as radically distanced – when not completely cut off – from the concerns of the young, discussion and energy were focused on life as it was lived at municipal level. Thus a succession of displacements – from the national to the local sphere – led to the formulation of a new social challenge hinging on municipal management.

Basically the conflict was between two rationales: that of the young Agora members with their total and utter commitment, and that of a municipality possessing all the legitimacy conferred by universal suffrage. For the municipality, participation was conceivable only within the framework of democratic dialogue: it was not the business of community associations to establish themselves as a political counter-force. City Hall's priority of putting an end to the stigma attached to the very name of Vaulx-en-Velin was approached as a globally municipal affair; this in itself was not incompatible with support measures for the most disadvantaged neighbourhoods, but at the same time the municipality had opted for focusing its efforts on the creation of facilities – outside the ZUP – that would boost the standing of Vaulx-en-Velin and its residents as a whole.

> The negative image problem that dogs Vaulx-en-Velin cannot be resolved by a simple rehabilitation of the neighbourhood as originally created ... What is needed are new centres of interest and thus new facilities comparable to those found in the central neighbourhoods of the Lyon metropolitan area.[13]

This agenda was given concrete expression by the construction, in a low-density area, of a central area, notably including a maternity hospital, a regional multicultural centre, a planetarium and a high school. The intention was to create a hub that would become a meeting point for all Vaulx-en-Velin residents and give them a sense of pride. Seen by the authorities as a response to the requirements of the population as a whole and as a means of boosting the municipality's image, this venture was perceived by the young members of Agora as the ultimate sign of political abandonment. For them municipal management only had meaning if it took as its point of departure the situation in the symbolic territorial entity of Mas du Taureau. In their view nothing the authorities did elsewhere was done with them in mind. Worse still, they felt that the creation of the city-centre facilities was happening at the direct expense of an improvement of their living environment.[14]

The socio-cultural centre – that barometer of life in the neighbourhood – was a focal point for tensions that burst into the open in a conflict between director and staff in February 1992. The director resigned in May, followed a few weeks later by the chairman, the treasurer and other members of the board. The centre was kept open by its team of coordinators, user representatives on the board – led by the Agora president – and young volunteers from the neighbourhood, and activities continued more or less as usual during the summer. For the young volunteers this was the chance to implement the principles they so fervently advocated. The Agora president's message was clear:

> We are no longer prepared to remain hostages of autocratic official bodies or consumers of an obsolete product. People in the social service system are annoyed by us, but the

---

[13] See Amendment No. 1 to the Vaulx-en-Velin DSQ (Neighbourhood Social Development) contract.

[14] This is illustrated by the discontent that crystallised around the planetarium, a project partly funded with money initially allocated to the Neighbourhood Social Development scheme under the terms of a January 1991 contract between the Rhône-Alpes Regional Council and the Municipality of Vaulx-en-Velin. The municipality obtained an amendment from the Region that allowed the abandonment or postponement of all projects not officially launched. The result was the annulation of three neighbourhood centres and a link road included in the contract and intended for different parts of the ZUP, notably Mas du Taureau. The uncommitted funds (in all 8.6m francs exclusive of tax) were then transferred to the planetarium project. Young people saw this, factually and symbolically, as extremely significant: public money originally intended to improve the living conditions of the most disadvantaged residents ended up being used for a facility for 'looking at the stars' and, what was more, built on a site with which they felt virtually no connection.

time for a challenge has come ... What we want are structures that foster independence, that give meaning to the notion of 'citizenship' the authorities keep going on about. Participation, sharing of responsibility, development of autonomy – that is our creed.[15]

When it realised it was losing control of the situation, the National Federation of Socio-cultural Centres decided to freeze funding, and the centre closed. Concerned by the loss of the main community facility in Mas du Taureau, but at the same time not wanting to give in to the challenge, the municipality faced the fact that there was no way of resolving the conflict. Gradually trashed by stone-throwers, the centre was destroyed in an arson attack in November 1994 and subsequently bulldozed: it lived on, however, as the symbol of a dialogue that was clearly impossible.

We see ourselves as entitled to criticize municipal policy, especially regarding the socio-cultural centre: going back many years now, this policy was created, with no local consultation, by incompetents with no knowledge of the terrain. So far City Hall has made no attempt to establish liaison between municipal departments and community associations ... Mr Mayor, we confirm our continuing readiness to work for true, ongoing local democracy.[16]

Discontentment was running so high it sometimes found expression in a determination to cut entirely free from the social service system. No longer believing in the possibility of a socially and politically negotiated improvement in their situation, some young people began talking in terms of a total break. At a time of especially acute frustration, this agenda found many supporters, especially among those with a long experience of ultimately futile commitment. Embittered by government policy on both Left and Right and by the failure of the Beur movement, they no longer had any real expectations of the authorities; and so accumulated disillusionment, not to mention purely personal considerations, took them towards an increasingly radical stance. Their analysis was in some cases extremely clear, well-argued, marked by a rigorous awareness of the political components of the situation – and totally devoid of any real hope: a point frequently made was that for young people from disadvantaged neighbourhoods, violence remained one of the very few feasible, effective means of getting any attention and, ultimately, any help from the authorities.

The Municipal Election Campaign of June 1995

Nonetheless the majority of Agora's young members stayed with a stance that, while openly conflictual, still sought outreach and partnership. This meant tough preconditions for cooperation with official bodies, but such cooperation was never excluded in advance. The ultimate victory of the anti-separatist line found what was doubtless its most significant expression in the decision to take part in the municipal elections of June 1995. This was an instructive episode in at least two ways: firstly, via

---

[15] Extract from an interview.
[16] Agora document.

the internal organisation of the campaign and the proposals made to the population, it provided a clearer idea of how the autonomy-for-residents was put into practice; and secondly, it enabled an evaluation of Agora's capacity to mount an authentic political project in a complex institutional context.

Agora as such did not run for election; rather it was the source of the 'Vaulx-en-Velin Chooses' ticket, an idea born and brought to maturity in the Association's offices. With Agora as the driving force, the idea rapidly attracted candidates from different backgrounds, the plan being to draw up an overall municipal programme appropriate to the election situation. Initial cooperation was with other neighbourhood associations with real roots in the ZUP and an overall approach based on resident participation. Many other people subsequently joined the movement, as candidates or not, the result being an astonishing mix of sympathisers or card-carrying members from the green, socialist and communist camps, and citizens with no particular political allegiance. All in all this was a highly tenuous grouping based mainly – and by default – on a global rejection of the municipality's record.

Given this relative hybridity, the campaign's organisational approach was designed to ensure expression and control by all concerned – including residents – while adopting a necessary division of labour and a codification of the rules. To this end, working groups were set up[17] involving thirty-three people in all. A campaign director was put in charge of day-to-day coordination of the groups and as often as necessary – on average twice a week – a campaign committee met to discuss their proposals. Comprising twenty-one people with no other organisational responsibility whatsoever – a means of ensuring independence of assessment – this committee was the sole decision-making body. Voting by show of hands required a two-thirds majority. A general meeting of all committee members and interested residents took place every two weeks, as a way of ensuring direct, regular interaction with the population. With a view to extending contact to the largest possible number of people, each meeting was held in a different venue, thus covering most of the municipality. The attention given to publicising these forums was a further indication of the importance accorded to dialogue with residents: posters everywhere invited people to attend and a minibus fitted with a loudspeaker toured the streets.

The general system turned out to be satisfactory: it proved viable and functional and provided a means of managing internal tensions by subjecting each member to scrutiny and control by the group as a whole. To ensure universal respect for the principles and procedures laid down in the rulebook, a six-person monitoring committee was set up – but never had to take action.

The ultimate shaping of the programme presented to the residents of Vaulx-en-Velin was the direct outcome of this cooperative approach, representing an ideological base acceptable to all concerned. Some of the young people, from Agora and other associations, had difficulty in accepting that subjects in which they had little interest – local taxes, for example – should be included. The character of the election and the influence of various partners not exclusively concerned with disadvantaged neighbourhoods resulted in a broadening of the programme, a more general emphasis

---

[17] In charge, respectively, of the programme, communication strategy and finance.

and a shift in terms of the issues. In the preliminary discussions the question of the neighbourhoods as such had hardly been raised: rather, subjects had been approached sector by sector.

The definitive programme embodied a dual rejection: firstly that of a Left/Right interpretation and stance, given that the conclusions reached and the measures advanced bore no relation to this kind of split; and secondly, of the least trace of community identity. There was not the faintest reference, direct or indirect, to the question of immigrants or religious backgrounds. For the rest, the campaign invariably focused on seven issues approached in an extremely down to earth manner: (a) making education and employment absolute priorities; (b) establishing real democracy and consulting Vaulx-en-Velin residents on all major issues; (c) using locally-oriented policy to improve the quality of life in the neighbourhoods in question; (d) increasing public safety and recreating a climate of trust; (e) supporting and promoting the community association movement; (f) reducing local taxes and improving the management of public funds; (g) transcending political divisions so as to work together for the good of Vaulx-en-Velin.

However, despite this undeniable determination to reach the population as a whole, the results obtained by the Vaulx-en-Velin Chooses ticket showed marked geographical and sociological differences. With 18 per cent in Mas du Taureau, it became the main opposition to the sitting municipal team, and earned lower but substantial scores in the neighbouring parts of the ZUP. At around 2 per cent the figures from the socially better-off areas were all but insignificant. The overall result was 7.23 per cent of all votes cast. Encouraging for some, disappointing for others, these statistics are open to different interpretations, but one thing is certain: on the very site of the riots of October 1990, a number of local actors had accepted – even if only momentarily – the challenge of commitment and political participation, and done so in a context in which many people were tempted by discouragement and/or mindless violence.

The situation in Mas du Taureau today gives little cause for optimism. The enthusiasm and effervescence of the municipal campaign have dropped away without any improvement in the plight of the residents. Worse still, the economic and social indicators all point to a downward trend. While not the cause of this state of affairs (Delarue 1991, pp.40-44), the 'politique de la ville' has shown its inability to provide responses appropriate to the type and extent of the problems involved.[18]

Yet with these difficult neighbourhoods looking like the epicentre of an overall reshaping of social issues, two questions remain. One has to do with the capacity of the authorities to halt an ongoing process of segregation and impoverishment leading, in some areas, to a concentration of manifestations of 'disaffiliation' (Castel 1995).[19] The other calls into question the way our institutions function, especially in terms of adapting to demands increasingly formulated in terms of autonomy. The issue here is

---

[18] Perhaps this is to be seen as the result of 'unreliability', 'improvisation', reliance on 'optimistic declarations' and ultimately lack of political will on the state's part (Le Galès, 1995).

[19] This situation marks a return to issues that surfaced c. 1830. In this sense 'there is a similarity of status between, for example, the 'useless to the world' vagabonds of the period prior to the Industrial Revolution and various categories of today's 'unemployables'" (p.16).

the possible emergence of a pluralist local democratic scene. Is it a good thing for democracy that ballot-box legitimacy means control not only of City Hall but also, by and large, of social welfare activity as a whole? At local level in France, political and community association-based challenges to established power structures are still embryonic. The Agora initiative and others like it are important in that they are proof of vitality and a degree of fresh political commitment, but their effects in terms of social change are limited: decentralisation might have considerably increased the power of city hall, but generally that of residents remains as restricted as before (Mabileau 1997).

The Formation of MIB (Immigration and Neighbourhood Movement)

Agora members have to cope not only with the everyday problems of the local social context, but also with an uncertain and extremely limited capacity for getting a political hearing. The association's remarkable staying power – thirteen years – and the lasting commitment of its most active members, not least its founder-president, provide an interesting illustration of the general evolution of its activity, especially over the last few years.

Born in a climate of virtual insurrection, Agora has gradually become a major actor on the local political scene. Historically, its development took place on the fringe of the usual participation mechanisms, beginning with a radical condemnation of parties and movements on both Left and Right. While solidly rooted in the neighbourhood and the municipality, its connections with the broader world are tentative and fragile. One of the questions now facing Agora is that of extension of its reach and level of influence: its local character is not an obstacle to expansion and more global action, although these can easily go unnoticed because they operate outside the classical mobilisation structures and as a rule are not taken up by the national media.

Even so, since the June 1995 elections Agora's network of links with other community associations all over France has continued to grow, culminating in the formation of MIB (Mouvement de l'Immigration et des Banlieues/Immigration and Neighbourhood Movement). As a means of self-affirmation in social, political and media terms, Agora is looking for a way out of its relative isolation via alliances with those who, especially in deprived areas, share more or less the same aspirations. Over recent years Agora has taken part in action against racism, discrimination and the 'double penalty'(imprisonment followed by deportation for migrant offenders), while declaring sympathy and support for the 'Sans-papiers' (immigrants with no papers), the homeless, and the Palestinian cause. Among its regular partners are the advocates of an alternative society of ATTAC (Association for taxation and financial transactions in support of citizens), the DAL right to housing group and the Revolutionary Communist League (LCR), whose agenda Agora has always more or less actively backed.

On the other hand, a new note has been sounded by Agora's official rapprochement with religious associations like the Young Muslims Union (UJM). In a memorandum dated 23 October 2000 and quoted in *Lyon Mag'* 2002, p.62, France's Renseignements Généraux intelligence service expressed its concern at the creation, on 25 September 2000, of Divercité-Agora, 'a body organised by Pierre-Didier Tché-

Tché Apea, founder-president of Agora in Vaulx-en-Velin; Abdelaziz Chaambi, UJM founder member; and Abdelmajid Mokeddem, founder-president of Interface.' This, according to the memorandum, confirmed 'the existence of a network of such associations in the Lyon metropolitan area, with Divercité-Agora as the non-religious counterpart of the Greater Lyon Collective of Muslim Associations initiated by the Young Muslims Union.'

All normative judgements aside, this kind of alliance is open to two complementary explanations. Strategically the rise of Islam in disadvantaged neighbourhoods offers Agora an unheard-of opportunity to broaden its field of action. And ideologically the Islamic revival fits closely, at least in some respects, with contemporary individualism: it cannot be reduced merely to an archaic sectarianism practised by a few bearded fanatics in the pay of foreign powers, for among the multiple components of this religious affirmation is a markedly modern emphasis on individual identity and autonomy. Thus Agora and the UJM have found common ground here. Seen in this light, Agora's behaviour is consistent with what the association has always wanted to do and the values it has always stood for. The fact that the religion-based associations use this alliance to further increase their influence in sensitive neighbourhoods – perhaps in the long term to the detriment of secular organisations like Agora – is another matter. The time for serious questioning as to the prerequisites for resident involvement in political and local affairs will come if and when the religious groups begin to function on the fringes of the associations and – worse – to challenge their rationale

Conclusion

France's sensitive neighbourhoods are a strikingly accurate mirror of social change, and this is particularly true in terms of two related situations: the rise in poverty rates and the spread of a cultural liberalism that emphasises individual participation and fulfilment. As a result of this dual shift, the forms of social, political and civic commitment that once worked in favour of the integration of immigrants now contribute to their exclusion. Insufficient application of the principles of 'democracy' and 'participation' that it advocates, means that the 'politique de la ville' fails to respond effectively to the aspirations of those seeking involvement in society. The riot is one of the radical manifestations of the anger and sense of abandonment felt by many residents of the areas in question. Nonetheless the Agora experience shows that certain young people do try to break out of the downward spiral by commitment to local participatory initiatives. Since its creation Agora has made autonomy of judgement and action its main objective, and in the course of some ten years has developed an analysis and a political agenda that were taken to their most advanced point during the municipal campaign of June 1995.

Commitment by young Agora members sprang not out of identitarian concerns – based, for example, on religious or ethnic loyalty – but out of dissatisfactions and demands rooted in modernity. Because they were not heard, the assertions of 'autonomy' are now being backed by the 'demands of Islam' that doubtless indicates a major shift in the life of the association.

References

Bell, D. (1979) *Les contradictions culturelles du capitalisme*, Paris, Presses Universitaires de France.

Bélorgey, J.-M. (1993). *Evaluer les politiques de la ville*, Paris, Comité d'évaluation de la politique de la ville.

Bonetti, M. (1994) *Le bricolage imaginaire de l'espace*, Paris, Desclée de Brouwer.

Castel, R. (1995) *Les métamorphoses de la question sociale. Une chronologie du salariat*, Paris, Fayard.

Chaline, C. (1997) *Les politiques de la ville*, Paris, Presses Universitaires de France.

Delarue, J.-M. (1991) *Banlieues en difficultés : la relégation*, Paris, Syros.

Donzelot, J. et P. Estèbe (1994) *L'Etat animateur : essai sur la politique de la ville*, Paris, Esprit, 1994.

Dubet, F. (1987) *La galère : jeunes en survie*, Paris, Fayard, 1987.

Dubet, F. (1994) *Sociologie de l'expérience*, Paris, Seuil.

Fontaine, J. (2003) La politique de la ville au risque de l'évaluation. Une politique ordinaire?, dans D. de Béchillon *et al.* (dir.), *Droit et politiques publiques*, Paris, LGDJ. pp.251-284.

Grafmeyer, Y. (1994) Regards sociologiques sur la ségrégation, dans J. Brun et C. Rhein (ed.), *La ségrégation dans la ville*, Paris, L'Harmattan, pp.85-118.

July, S. (1990) Le modèle Vaulx-en-Velin, *Lyon-Libération*, 13 et 14 octobre, p.2.

Lapeyronnie, D. (1993) *L'individu et les minorités : la France et la Grande-Bretagne face à leurs immigrés*, Paris, Presses Universitaires de la France.

Le Galès, P. (1995) Politique de la ville en France et en Grande-Bretagne : volontarisme et ambiguïté de l'Etat, *Sociologie du travail*, Vol.37, No.2, 1995, pp.249-271.

Lyon Mag' (2000) Islamistes dans les banlieues lyonnaises, une note inquiétante des R.G., No.112, pp.60-2.

Mabileau, A. (1997) Les génies invisibles du local. Faux-semblants et dynamiques de la décentralisation, *Revue française de science politique*, vol. 47, No.3-4, juin-août, pp.340-376.

Marie, J.L., P. Dujardin et R. Balme (2002) *L'ordinaire. Mode d'accès et pertinence pour les sciences sociales et humaines*, Paris, L'Harmattan.

Pharo, P. (1990) Les conditions de légitimité des actions publiques, *Revue française de sociologie*, No.31.

Sardais, C. (1990) Rapport sur la mise en œuvre de la politique de la ville.

Sayad, A. (1992) Qu'est-ce qu'un immigré?, dans *L'immigration ou les paradoxes de l'altérité*, Bruxelles, De Boeck, pp.51-77.

Sayad, A. (1977) Les trois « âges » de l'émigration algérienne en France, *Actes de la recherche en sciences sociales*, No.15, juin, pp.59-79.

Schnapper, D. (1986) Modernité et acculturations. A propos des travailleurs émigrés, *Communications*, No.43, pp.141-168

Schudson, M. (1994) La culture et l'intégration des sociétés nationales, *Revue internationale des sciences sociales*, No.139, février, pp.79-100.

Tapinos, G. (1988) Pour une introduction au débat contemporain, dans Y. Lequin (dir.), *La mosaïque France. Histoire des étrangers et de l'immigration*, Paris, Larousse, pp.429-447.

Taylor, C. (1994) *Le malaise de la modernité*, Paris, Le Cerf.

Touraine, A. (1994) *Qu'est-ce que la démocratie?*, Paris, Fayard.

Chapter 13

Instituting Proximity: The *Arrondissements* of Paris, Marseille and Lyon Since 1983[1]

Melody Houk

Like its European neighbours, France has been engaging over the last twenty years in fairly intense institutional activity to reorganise local political and administrative functioning. Increased official responsibilities for local government, the increased importance of the administrative unit called the *région*, unprecedented development of relations among municipalities and simplified procedures for intercommunal cooperation, the development of citizen participation structures at the infra-communal scale – all these phenomena attest to an ongoing attempt to develop efficient, democratic local organisation based on a harmonious distribution of jurisdiction and official responsibilities. A number of recent political science and sociology studies have focused on – and helped fuel – the increase in the power of local government institutions.[2] These studies, whether formulated in terms of locality and territoriality, public action, public policies, governance work to show what is being constructed over time and through institutional and political struggles and actors' strategies, by efforts at the national and European Union scales to rationalise political-administrative organisation.

The present study, focused on the emergence of the infra-municipal echelon in France's three largest cities, Paris, Marseille, and Lyon, is in keeping with this general approach. I look at the somewhat hybrid institution that was legislatively created in the early 1980s for these three cities: *arrondissement* councils and mayors. The *arrondissement* is an old administrative sub-unit of these three cities which have developed through inclusion of surrounding boroughs until the beginning of the 20th century. Paris is divided in 20 *arrondissements*, Lyon has nine and Marseille a total of

---

[1] The present analysis is based on empirical research (interviews, documentary study, observations) conducted in the 20th and 15th *arrondissements* of Paris, the 5th and 8th *arrondissements* of Lyon, and the 1st and 8th sectors of Marseille from 1998 to 2001. It is part of a doctoral thesis I am writing at the Centre de Sociologie des Organisations in Paris (CNRS/Institut d'Etudes Politiques de Paris) under the direction of Olivier Borraz; my grateful thanks to him for his judicious comments on earlier versions of this text.

[2] On the strengthening of the regional echelon, see Négrier and Jouve 1998, and Nay 1997; on renewing/new forms of intercommunal cooperation see Caillosse 1994, and Baraize and Négrier 2001; on municipal government see Borraz 1996, and Qui gouverne les villes ? 2000; on cities as actors in urban governance see Bagnasco and Le Galès 1997, Le Galès 2002 and Sellers 2002.

sixteen. I will here use the term of 'district' for convenience, even though these French districts are not comparable to city districts one can find in Chicago or London for instance as the official role of today's district councils is to advise the greater city council. Because *arrondissement* councils are now directly elected by citizen residents, they have real political legitimacy. But designed as an advisory body, they have no decision-making or direct action-taking power in *arrondissement* affairs. This type of institutional structure stands in contrast to the regional echelon and intercommunal structures, both of which involve strong official responsibilities and, for the latter, weak though not entirely non-existent democratic legitimacy. It therefore raises different issues, including the question of how such an institution and the elected representatives engaged in running it manage to exist without the traditional attributes of power of modern Western democracies: decision-making power and financial resources. Before considering this question, it is useful to explain the infra-municipal arrangement created by the law and the context in which it was conceived.

The law on the administrative organisation of Paris, Marseille, and Lyon known as the 'Loi PML' was passed by the French parliament on December 31, 1982. Behind it was an ambitious and intensely political plan. Socialist President François Mitterrand was intending to destabilise the partisan system of his political adversary Jacques Chirac, then mayor of Paris and leader of the opposition Rassemblement pour la République (RPR) party on the right, by carving up the capital into 20 full-fledged *communes*. The plan was abandoned after Chirac's successful 'battle of Paris', a full-blooded campaign against the project, based on a systematic denunciation in the media of the political move that was behind the government's project and the absence of consultative process regarding such an important matter. The government stepped back and withdrew the initial project, proposing a reform of the political-administrative organisation of not only Paris, but also Marseille and Lyon, the 2 biggest cities following the capital, and that would not break the city's unity. Then the state government's position was actually quite seriously undermined by the fact that the Minister of the Interior and Decentralisation in charge of executing the plan was none other than Gaston Defferre, mayor of Marseille since the early 1950s and none too eager to share his power with any infra-municipal authority. What was left of the original project was a hybrid, timid, arrangement that only appeared to decentralise power.

The 1982 law endowed infra-municipal districts (*arrondissements*) of Paris, Marseille, and Lyon with councils elected by universal suffrage and presided over by a mayor (*maire*). Their official role was to advise the city council on all district matters. Decision-making and financial resources thus remained in the hands of the central city executive; district halls (*mairies d'arrondissement*) would not have their own budget or run their own services. The councils were granted the responsibility of managing a number of neighbourhood facilities: 'green spaces' of less than one hectare, crèches, public baths, youth centres, gymnasiums and physical education playing fields, and 'all equivalent facilities ... that are primarily for the use of *arrondissement* inhabitants' (Article L2511-16 of the Code Général des Collectivités Territoriales or CGCT). The list reflects the timidity of the reform. Running these facilities involved ensuring their daily functioning and maintenance, for which district authorities were to receive a small subsidy, whereas personnel management remained the official responsibility of the central city government, as did all decision making on and investment in facilities. District mayors and councillors would not be able to build facilities on their territory or implement policies they had declared support for in their election campaigns. The institution created by the Loi PML was then sharply criticised for its weakness, and few politicians thought it would last.

Twenty years later, however, district authorities are still here and have become increasingly visible, while their official responsibilities and financial means derive from the same legislative source. How have they made a significant place for themselves in the landscape and management of the three cities? This question is our focus throughout the present chapter.

Overall, this paper aims at casting light on how the infra-municipal level has emerged through the particular action repertoires used by local representatives since the 'Loi PML' went into effect. These repertoires evolved over time, and may be said to fall into two main periods. The first, which is the focus of the first part of the chapter, stretched from 1983 to 1995, the first two terms of office the law was in effect. During this phase, the legal texts were implemented – that is, district councils were set up with a mayor at their head – in a relatively hostile political environment, and the infra-municipal echelon was essentially a relay for the municipal one, remaining fairly undifferentiated from central city government. The 1995 municipal elections marked a turning point, the beginning of a significant change in action repertoires and the way the infra-municipal echelon positioned itself. In the second part of the chapter I examine how, with the arrival of significant numbers of district mayors from the political camp opposed to that in power in central city governments, a dynamic of differentiation took over. The new representatives set out to organise and structure their territory and create power relations in which they could have a distinct existence and win city hall recognition of the *arrondissement* as a full-fledged institution within municipal government.

Implementing The 'Loi PML': The Difficult Emergence of the Infra-Municipal Echelon

The 1983 city elections were the occasion for implementing the new 'Loi PML', but this was done under the aegis of city mayors more or less openly opposed to it, not at all convinced of the relevance of the new arrangement. In all three cities, the central executive power in charge of implementing the law worked to minimise its impact on existing governing modes, and in general to absorb the new arrangement into itself. Jacques Chirac was already pledging while the law was being drafted to get rid of it as soon as possible and promised he would interpret the text as restrictively as possible in the meantime. For his part, Gaston Defferre did not want the infra-municipal echelon, now politically representative and endowed with its own institutional existence, to emerge as a government entity threatening his own authority. The arrival of district mayors, chosen by and from among duly elected district council members, raised fears that a form of counter-power would be erected; accordingly, the central executive powers sought to interpret the text in minimalist terms. The impulse to absorb the reform was strengthened by the results of the city elections in Paris and Lyon: the right won all *arrondissements* in the two cities.[3] This configuration made calls for full application of the law unlikely. In this context, the newly elected district officials struggled to exist and win legitimacy through two types of action: imitating the city council, and positioning themselves as a relay between the city residents and city hall.

The Games and Stakes of Imitation: 'Cardboard' District Mairies?

The *arrondissement*-level arrangement created by the 'Loi PML' for Paris, Marseille, and Lyon took on the shape and features of the greater municipal institution. It was, and still is, composed of a council that elects a mayor from among its members and votes on issues that concern its territory. The mayor presides over the council, assembles a team of deputies, and sets up a cabinet. The municipal referent is present semantically and infra-municipal echelon actors immediately adopted city hall as the relevant frame of reference. Construction of the *arrondissement* institution thus involved a mimetic dynamic where the model to be imitated was the republican municipal institution.

The legislative text itself was an essential vector for this, partially modelling the way actors, particularly the newly elected *arrondissement* representatives, invested their roles. During preliminary parliamentary work on the law, opponents to it had already expressed concern at use of the terms *maire* (mayor) and *adjoint* (deputy), seeing them as a source of ambiguity and encouragement for unwarranted political behaviour. And indeed, once *arrondissement* mayors formally took office, they played on this ambiguity between their title and their slight real powers. Actors' behaviour thus fuelled the mimetic process already latent in the legal text.

Mayoral cabinet, general secretary, the function of Public Records Office [district halls are in charge of keeping birth, marriage, and death registers], regular hours during which inhabitants can consult with representatives, council meeting hall – the citizen who walks unaware into an *arrondissement mairie* finds all the features of this echelon in the image of city hall. In terms of internal organisation and functioning, the local institution also reproduces to large degree the traditional municipal model. At the start of a term in office, *arrondissement* council members divide into groups along political lines, groups often modelled directly on those of the city council, each of which then designates a president. The district mayor meets regularly with his or her team of deputies (in full or restrained committee depending on circumstances and the particularities of the district council coalition arrangement) to examine projects on the council agenda. One may also find staff in charge of such areas as communicating information to the public or city planning, while the cabinet director is often given the job of public relations, i.e. receiving and responding to residents, businesses, and association or advocacy group representatives that come to lobby the mayor. Various theme-based commissions (e.g. 'City planning', 'Young children', 'Cycling') and ad hoc

---

[3] In Marseille, however, Defferre was not able to contain local opposition. The right won two out of six sectors, and the mayor was forced not only to grant a third to his long-term adversaries the communists, but also to form a coalition with them (though given his vast experience managing Marseille's political fiefdoms, this type of situation was not entirely new to him).

commissions (ZAC projects [Zones d'Aménagement Concerté: specially designated housing and public facilities development zones], specific urban development or rehabilitation projects, etc.) attest to the political will in district *mairies* to tackle local issues and involve local actors, in the same way traditional city executives do it through extra-municipal commissions. The areas of delegated responsibility and their designations also reveal a mimetic dynamic. There have been district deputies for 'Commerce, small and medium-size businesses, crafts', 'Employment and solidarity', 'Housing', 'The handicapped in the city', even 'Women's rights'. Though many of these areas do not fall within official *arrondissement* competence they are nonetheless indicated as deputy responsibilities. The use of such designations also attests, as is generally the case in city government teams, to local representatives' desire for political visibility and official engagement to give priority to specific/sensitive sectors or problems of public policy (Borraz 1995). As the Loi PML went into effect, local representatives showed a strong propensity to adopt the 'prescribed' forms of the municipal model, and this early interpretation has had an enduring effect on the institution.

The mimetic dynamic has been relying on two main mechanisms. First, for representatives called upon to give shape and substance to new functions, the municipal model was an obvious reference. Mode of election, titles, the representative function, etc. were all features that made them fundamentally similar to city representatives and guided role interpretation. Second, and parallel to this autonomous and somehow passive process, was the fact that district officials were anxious to establish their legitimacy, and they saw imitation of the traditional municipal institution as the best means of doing so, of endowing the newly created institution – as if by transitivity – with the same kind of legitimacy enjoyed by the municipal referents *maire* and *mairie*. On this point I follow the neo-institutionalist analysis holding that an organisation tends to adopt the institutional forms that appear legitimate in the field it operates in.[4] This mechanism is a source of legitimacy for the institution itself and increases its chances of survival (Meyer and Rowan 1977). District mayors and mayoral teams in the three French cities thus strove, and still strive, to show that the infra-municipal institution has the same features as a full-fledged city government. They seek occasions (in addition to official ceremonies such as weddings and various inaugurations at which elected representatives wear the tricolour sash) to appear and intervene publicly, as these offer an opportunity to represent the *arrondissement* institution as highly similar to a classic city hall. In doing so they play on the confusion in the citizen's mind between appearance – the institutional façade as it were – and effective exercise of specific official responsibilities. The point for *arrondissement* officials is to 'act as if' they had initiated this or that investment – the long-awaited crèche, a junior high school extension, rehabilitation of this or that block of insanitary or otherwise unsafe buildings. The intention is as much to give the appearance of real decision-making power in the city as to actually have that power.

In general, this kind of *représentation* in the sense of theatrical staging was, and is, aimed at maintaining a façade of legitimacy in citizens' eyes. The municipal institution

---

[4] See Meyer and Rowan's analyses of institutional myths and mimetism (1977) and DiMaggio and Powell's study of institutional isomorphism (1983).

is a familiar structure in the French institutional landscape and the figure of the mayor enjoys a high level or 'capital' of citizen 'trust.' These are precious resources from which *arrondissement* representatives have sought to benefit from the beginning.

But legitimation through imitation could only take the emerging institution so far; the limits of the process were quickly reached. City hall's legitimacy derives of course from its ability to represent the interests of a majority of inhabitants (the municipal assembly is directly elected by the voters) and above all to deliver – decisions, facilities, housing, etc. Its specificity resides in its position at the intersection of three different types of logic, those of politics, the production of goods and services, and the locality or territory (Lorrain 1991). Its legitimacy thus derives as much from its ability to 'regulate everyday life' as to integrate these three areas. And it is precisely this 'daily regulation' dimension that is missing from the infra-municipal echelon as defined by the 'Loi PML', for while *arrondissements* can integrate the dynamics of politics and the locality, they cannot themselves produce urban goods and services or insure the type of 'everyday regulation' that gives the traditional municipal institution an essential part of its legitimacy and visibility. This is of course a serious limitation, and one that moved *arrondissement* representatives to seek and cultivate other sources of legitimacy.

New Connections Between City Hall and City Residents

After the 1983 municipal elections, district mayors and councillors could only occupy the limited space allotted to them by the law itself and the city executive's minimalist reading of it. The way they played their roles within this stringent framework was structured by two main factors: the district council's official, formal function of advisory body, and the on-the-ground territorial presence of these particular representatives, their direct contact with citizens and local realities. The combination of these factors in an environment that was, as we know, relatively hostile to the reform led the infra-municipal echelon, or more exactly its representatives, to position itself as a kind of conveyor belt between city hall and district residents.

Consulted by the municipal executive on the various projects directly touching their districts, both in the framework of the formal procedure specified by the law and the more informal relations among representatives of the majority, local representatives were led to develop a network of relations with the economic and social forces of their *arrondissement*. Contacts with association or advocacy group representatives, businesses, condominium managers and the like enabled district mayors and deputies to get an idea of local perceptions and reactions to city hall projects. In Marseille and Lyon this was no small feat. *Arrondissement* officials in those cities were in competition with representatives of already existing neighbourhood committees (Comités d'Intérêt de Quartier or CIQs in Marseille, Comités d'Intérêts Locaux or CILs in Lyon) which had been the favoured interlocutors of city hall for decades (Mattina 1999). The municipal majority knew very well how to choose between an 'ordinary' district councillor, even though he or she was endowed with electoral legitimacy, and a CIL or CIQ president who could boast of 'representing' or at least having direct access to several dozen resident voters. Likewise, the neighbourhood 'Comités' and associations had no doubt where real decision-making power and means for action lay.

Anxious to respond to questions from the local population on municipal projects, district officials were – and still are – on the lookout for information on planned action by the municipal government, information which officials at that level, technical service department heads as well as deputies, were not always in a hurry to communicate to them. They needed to establish a network of useful contacts within the municipal apparatus, in both executive and technical service departments. In Paris, *arrondissement* mayors and their main deputies were able to get closer to city hall from the start due to localised organisation of urban services (present in *arrondissements* in the form of local sections, boards, offices, etc.). Daily collaboration with local service offices gradually became a standard part of *arrondissement* officials' practice, one which took even deeper root after the 1995 elections. Such collaboration seems to have worked in the interests of district officials and local-level functionaries alike. The first group were looking for information and ways to intervene in district daily life matters, to participate in defining and handling local problems, while the second found local officials to be informative interlocutors who might well be ready to go up to city hall and request or demand further resources for one or another municipal policy area. In Lyon, it was possible to establish direct relations with services at the *communauté urbaine* level without going through municipal-level channels.[5] In general, the desire for effective proximity, i.e. a concern to take into account not only local realities but the significant fact that *arrondissement* mayors belonged to the city hall political majority, facilitated information exchange and the adapting of municipal action to local needs at the margins of the system (modifications in small planned changes to roads or green spaces, public street lighting, etc. i.e. changes already planned for in the city budget).

Another aspect of the job of local representative, one in which these 'new' office holders began investing themselves intensely, was handling residents'/citizens' demands. This was facilitated by the simple increase in number of local representatives. In establishing permanent political representation at the infra-municipal level, the Loi PML increased the total number of municipal officials and created a new category, closer to citizens than city councillors. It is not clear how much progress this really represents given that *arrondissement* councillors remain legally separate from the city council[6] and that certain districts or sectors are equal in population and surface area to good-sized cities.[7] Nonetheless, the number of representatives potentially present at the local level and directly representing the population more than tripled after 1983, going from approximately 100 to more than 500 in Paris, and from 60 to 200 in Marseille and Lyon. The number of municipal councillors alone (not counting *arrondissement* councillors) has also increased in the

---

[5] Lyon is one of France's 14 *communautés urbaines*, defined as a greater metropolitan area of at least 500 000 inhabitants including a city of at least 50,000.

[6] *Arrondissement* councillors and municipal councillors are elected simultaneously from a single list. One third of the *arrondissement* council is made up of municipal councillors, who also sit on the main council; the remaining two thirds serve only on the *arrondissement* council.

[7] The 15th *arrondissement* in Paris has nearly 225,000 inhabitants, a population comparable to Bordeaux.

three cities: the Conseil de Paris now has 163 elected members instead of 109 for 2,200,000 inhabitants (population in 1977), the equivalent of one official for every 20,000 inhabitants; the city council of Marseille went from 63 to 101; the city council of Lyon from 61 to 73.

It should also be stressed that the law created two categories of representatives: city councillors, called *grands élus* when they also hold a seat in parliament or serve as city hall deputies, and *arrondissement* councillors, whose range of action is limited as we know to the infra-municipal echelon. In fact, once elected some city councillors hardly set foot in *arrondissement mairies* and were rarely to be found on their council benches, and some *arrondissement* councillors (namely members of the opposition at the *arrondissement* level or in the majority but without particular responsibilities) did no more than attend council meetings. During the 1989-1995 term in Marseille, *arrondissement* council meetings were regularly rescheduled for lack of a quorum (Olive 1997, pp.212-213). Meanwhile, district councillors in the three cities sometimes became specialised in matters touching inhabitants' daily lives, an area that *grands élus* and professional politicians tended to neglect.

Lastly, one should pay attention to the problem of geographical disparities among intra-city districts, and therefore among their citizens. Inhabitants of smaller districts are in very close physical proximity to the district hall and institution, whereas in large, poorly served *arrondissements* or sectors, the hall itself is far away and hard to reach for many inhabitants. This is especially true for neighbourhoods in the north of Marseille.[8]

Because of their proximity, *arrondissement* officials rapidly became the first to receive and collect residents' petitions, claims, complaints, requests and demands of all sorts, from housing to job placement assistance not to mention matters of parking, school enrolment, and neighbour relations. That this should be so reflects not only citizens' confusion about the official jurisdiction and responsibilities of each government or administrative level (city hall, *département* council, regional council, police department, education authority, etc.) but also the ambiguous nature of their relation to political power. While this situation is characteristic of all political institutions in France, *arrondissement* officials with no other office turn out to be particularly ill-equipped to handle it given that they have no financial means or decision-making power. Logically, their role mainly consists in either transmitting citizens' requests and demands and redirecting individuals to the relevant service or, for those who hold a higher office, handling such demands directly.

When *arrondissement* officials cannot handle petitions or demands directly because they fall within city government jurisdiction, they mainly play the role of conveying

---

[8] In 1982, the legislator designed a specific frame for Marseille, instituting the sector (the addition of at least 2 *arrondissements*) as the territorial unit for the implementation of the PML reform. For electoral reasons mainly, Marseille was divided into highly unequal electoral sectors ranging from 70,000 and 245,000 inhabitants, each comprising one to four *arrondissements*. This move was roundly denounced by the national opposition as electoral *'charcutage'* aimed at ensuring Defferre's re-election. The mayor was indeed re-elected - without an all-city majority. When the right came to national power in 1986, Marseille's electoral districts were redrawn and sector size regularised, though political considerations were operative in this change as well. At present Marseille is composed of 8 sectors of 2 *arrondissements* each.

them toward the municipal apparatus. Still, every representative tries to make some headway in handling such appeals, to demonstrate to the citizen or association that has appealed to them that s/he can have some effect in fixing the problem, obtaining requested aid, arbitrating in favour of one rather than another solution, etc. While ability to provide access to central city government decision-makers or those who influence decision-making constitutes a major resource for these representatives, who are as we know themselves without decision-making authority and means of action, their ability to construct an image of themselves as 'officials with power' and to inform inhabitants on their action turns out to be just as important politically. And this ability is largely a function of individual skill (Garraud 1989, p.167).

There are, however, cases of officials with direct or indirect personal resources that enable them to handle some of the requests that citizens address to them as local representatives. Such resources have had a considerable impact in orienting and positioning the *arrondissement* institution in its environment, and on the perception that citizens have come to have of it. *Arrondissement* mayors in particular are not all 'equal before their function'. While belonging to the city government political majority can enable one to have more direct access to the centre, holding several offices can prove decisive when it comes to real or projected ability to intervene in favour of local 'interests'. A district mayor who is also president of an HLM bureau [Habitation à Loyers Modérés: moderate-income public housing] can put his/her cabinet directly in charge of handling some of the housing requests addressed to the *arrondissement mairie*. A district mayor who is also president of the *département* council can enable 'his or her' inhabitants to benefit from *département* resources, partially transforming the *arrondissement mairie* into a social service window. The infra-municipal echelon is thus fertile ground for the development of clientele-type relations between representatives and voter-citizens. The functioning of sector halls in Marseille is often approached in these terms by political science researchers. In some respects, *arrondissement* hall activity is disconnected from official legal duties and is determined instead by what individual representatives bring to it.

During the first phase of *arrondissement* existence under the PML law, spanning the first two office terms in which the 'Loi PML' was in effect (1983-1995), the infra-municipal echelon was largely characterised by integration into the municipal apparatus, and local officials' dependence on the central level with regard to what points went on the agenda, financial means, action, and time-tables. *Arrondissement* officials tried to position themselves as relays between inhabitants and central city government. Some used the political legitimacy of their office to mobilise resources external to it, thereby constructing a certain capacity for handling social needs. Along with Olive's analysis for Marseille, we deem it relevant to emphasise the diversity of configurations that emerged when the law was first implemented, and of the types of action logic used by officials whose resources and interests varied (Olive 1997, pp.215-222). In fact, how the *arrondissement* level functioned during this period depended most on the individual representatives; individual resources (access to or influence over the municipal executive, holding more than the one office) were at a premium. This situation was perfectly consistent with city mayors' efforts to keep the *arrondissement* from emerging as a full-fledged institution, a collective counter-power. Lastly, *arrondissement* representatives' actions during this period worked to strengthen the legitimacy and regulatory power of the central city government, rather than those of the infra-municipal echelon. However, the change in political configuration that resulted from the 1995 city elections led to a repositioning of *arrondissements*, produced by a dynamic of differentiation that was strongest in *arrondissements* that had gone over to the opposition.

The Period Following the 1995 Elections is Characterised by a Dynamic of Differentiation and the Gradual Institutionalisation of the Infra-Municipal Echelon

The 1995 city elections marked a turning point in how the Loi PML was applied. In addition to major political shake-ups – six out of 20 *arrondissements* in Paris and three out of nine *arrondissements* in Lyon went to the left; Jacques Chirac moved from Paris' city hall to the office of President of the Republic; Marseille elected a mayor from the centre-right UDF party (Union pour la Démocratie Française) – these elections gave rise to new action repertoires at the infra-municipal echelon. Implemented above all by district mayors in the opposite political camp from the one in power in central city government, these repertoires have tended to complement rather than replace those used previously. They followed two major lines: constructing institutional relations with the local territory, and initiating or keeping up a power struggle between the *arrondissement* echelon and central city government. More generally, they indicate a dynamic of differentiation from the central municipal apparatus, and their effect has been to institutionalise the infra-municipal echelon.

Using the Institution and the Locality to Differentiate Arrondissement-Level Action

After the 1995 city elections, a significant number of district mayors found themselves, as political opponents to the central executive – for the first time in Paris and Lyon – having to exist and construct legitimacy without their predecessors' special access to central city government.[9] In response to this situation, local representatives turned back to the locality, *le territoire*, choosing to develop the institutional existence of the *arrondissement* authority on *arrondissement* territory, independently of central city government, by creating instruments and bodies specific to the infra-municipal echelon, particularly in the area of citizen participation. This choice led to a process by which institution and locality mutually constructed each other, a dynamic in which a form of proximity specific to the infra-municipal echelon was created.

When new district teams took office they immediately began outreach to inhabitants and associations, advocacy groups, and socio-occupational actors present on their territory, to get a clear idea of existing information and opinion channels and

---

[9] In Paris for instance, opposition district teams had to deal with the sudden closing down of access to city hall, where the city services received instructions immediately after the elections not to deal directly with opposition *mairies*, and central bureaus were told to report to the mayor's cabinet on any contact with left-led *arrondissements*. But this behavior was more a reaction to the political shake-up than a real *containment* strategy on the part of the municipal executive, and the impasse it produced was gradually overcome by the pragmatic logic of cooperation operative on the ground.

develop new ones. As mentioned, *arrondissement* authorities already tended to try to position themselves as non-circumventable intermediaries between the population and central city government able to legitimately represent local interests to the central executive. The newly elected teams followed this same logic, with the difference that certain mayors from the left also worked to institutionalise those relations with 'their' territory.

Michel Charzat, socialist mayor of the 20th *arrondissement* in Paris, is a case in point. He developed two new political practices in the area of citizen participation: close consultation with residents and transparency. After a political campaign focused on just these concerns, the new mayoral team set up participatory bodies and transparency commissions, thereby actualising within the *arrondissement* framework some values and arrangements usually operative in the traditional municipal framework at the city or town level. A modified *Comité de Consultation et d'Initiative d'Arrondissement* (CICA)[10] was launched, providing a more active role for associations; seven neighbourhood councils were created, covering the 20th *arrondissement* territory in its entirety, together with a children's district council; a procedure was implemented allowing citizens to submit points for district council agenda through local referendum; transparency commissions were set up on attribution of *crèche* slots and proposals for attribution of public housing. This panoply of new instruments gave substance to the *arrondissement mairie*, making it 'thicker' and more visible as it 'stepped down' to meet the territory. Other leftist mayors and their teams developed other experiments and arrangements, not provided for by the law, under the heading 'participatory democracy': a forum for associations in the 18th *arrondissement* of Paris, neighbourhood councils in the 19th and later 10th *arrondissements* as well as in Lyon's 8th. In Marseille, meanwhile, the newly elected communist district mayor, Guy Hermier, set up neighbourhood CICAs, a territory-based form of the arrangement created by the law.

Opposition mayors also worked to develop their own communication tools, since the possibilities offered by local newspapers were often quite limited. Information bulletins on association activities and newspapers published by *arrondissement* authorities began to multiply: in Paris *La Gazette du 20ème*, *Le Journal du 18ème*, and for the 11th *arrondissement Bastille Nation République*; in Marseille *Quartiers Nord*, published by the authority of the 8th district (15th and 16th *arrondissements*), which also designed its own logo. Local representatives used such spaces to make visible both the actions of district *mairies* and the life of the locality, thus cultivating local identities and even helping construct them.

Central city governments did not approve of these initiatives, and they made a point of calling district mayors to order on the limited possibilities available to them under the Loi PML. The new mayor of Paris, Jean Tiberi, opposed the creation of

---

[10] The law provided for each *arrondissement* to set up a CICA that would bring together representatives from interested local associations and called on the CICA to participate at least once a quarter in *arrondissement* council debates. Its members were to act in an advisory capacity and could present any and all issues or initiatives related to their activity in the *arrondissement* (Article L2511-24 of the CGCT). But given the fact that *arrondissement* authorities had virtually no resources or jurisdiction, this arrangement for enabling associations to 'participate' in local affairs had proved a failure.

neighbourhood councils proposed by 20th *arrondissement* mayor Michel Charzat, arguing that *arrondissements* were not local authorities and therefore did not have the same right as the city council to create extra-municipal commissions. After threatening Charzat with a lawsuit, Tiberi finally settled for refusing to officially recognise neighbourhood councils. In Lyon Raymond Barre contested the neighbourhood councils created by the socialist mayor of the 8th *arrondissement*, Jean-Louis Touraine. At issue here above and beyond legal questions was the access that municipal and infra-municipal echelons had to the citizen and more broadly the ability of each to construct its own kind of proximity. The cases of Lyon and Marseille are particularly revealing. Whereas in Paris, Michel Charzat could position himself as a pioneer of local democracy, in the other two cities, the initiatives of the *arrondissement* and district mayors were consistent with previous practices. These cities already had their CIQs or CILs, attesting to a conception of proximity going back to the late nineteenth century and strongly rooted in the institutional landscape, where they structured relations between municipality and resident population.[11] The participatory and representative bodies for neighbourhood inhabitants created by Jean-Louis Touraine in Lyon and Guy Hermier in Marseille clearly reflect a concern to equip the territory with their own specific links and the means for consultation and mobilisation of the local population. In these areas, the infra-municipal echelon was entering into competition with municipal government.

Central city government resistance to opposition *arrondissement* mayors' initiatives hardly proved a real obstacle, however. Because that resistance received a great deal of media coverage, it actually gave *arrondissement* mayors extra visibility on the local stage. Indeed, a strategic strong point of this particular action repertoire was to attract media coverage while remaining extremely difficult to attack. A mayor such as M. Charzat knew how to attract and capitalise on press interest in the infra-municipal echelon. He made a point of communicating fully and openly on his policies and making the 20th *arrondissement* appear a 'laboratory of local democracy in Paris'.[12]

Above and beyond getting media coverage, the move by opposition mayors to set up citizen participation arrangements served two purposes. First, formal arrangements of the sort were a means of accessing the opinions of actors concerned by projects sent down from city hall, projects that the *arrondissement* council would have to pronounce on. The charter of 20th *arrondissement* neighbourhood councils specified that *arrondissement* council recommendations on projects submitted by city hall had to take into account the position of the particular neighbourhood council consulted. Given the status of *arrondissement mairies*, however, it is incorrect to

---

[11] At the origin of these late-nineteenth century bodies were groups of citizen volunteers mobilised to petition the public authorities to provide neighborhood facilities such as sewers, local road maintenance, streetlighting, electrification, etc. For a detailed, insightful history of Lyon's CILs, see Joliveau 1987.

[12] A year after Charzat was chosen to lead the 20th *arrondissement*, the leftist daily *Libération* - only one of the newspapers providing detailed coverage of initiatives there - published an article entitled 'Charzat, or lab experiments in the 20th' (5 June 1996). With the arrival of leftist councilors in Paris and Lyon, the figure of 'the *arrondissement* mayor' became the focus of regular news reports and individual portrait pieces, which in turn gave it a certain depth and sustained as well as reflected infra-municipal echelon visibility.

consider these experiments involve direct citizen participation in decision making. Neighbourhood councils, for example, were conceived as advisory bodies for *arrondissement* councils, themselves endowed with no more than an advisory role with regard to the city council. The limitations of such initiatives are strong (Houk 1996, Blondiaux and Lévêque 1998). But the ad hoc task commissions set up within neighbourhood councils did enable local officials to give informed recommendations and above all to show that they could improve on city hall projects, making them more amenable to *arrondissement* interests. In this respect, and especially given that *arrondissement* halls do not have their own technical services, it can be said that these arrangements helped adapt city government projects to local needs.[13] Neighbourhood councils are indeed a source of local expertise for *arrondissement* representatives (intervention of associations, professionals such as architects and social workers living in the neighbourhood, etc.), and the good will of ordinary inhabitants, even though they tend to get involved in single issues, can also be of appreciable assistance (in such activities as gathering information and opinions from neighbours, counting parked cars, directly observing behaviour of public space users, etc.).

Moreover, neighbourhood councils and other forms of participatory democracy have proved to be a source of legitimacy for opposition mayors. This is true at the district level, where they strengthen mayor's visibility and legitimacy as representative of inhabitants and other local interests, but also in the municipal apparatus, where they give district council recommendations and positions a legitimacy that may be described as democratic and that directly benefits *arrondissement* mayors in their relations and negotiations with the municipal executive.[14] Such legitimacy, particular to local structures, is, as Dominique Lorrain explains, 'founded above all on their ability to 'know' and satisfy needs … [T]he reference to 'legitimate' need satisfaction is what makes it possible to justify new projects or local tax increases to inhabitants. And it is this knowledge of needs that founds local representatives' legitimacy in the eyes of the central state. In other words, they are recognised because they 'speak of social needs' in the name of the population they represent' (Lorrain 1981, p.16). Though in the case of *arrondissements* there is no direct satisfaction of needs, a parallel can be drawn between the legitimacy obtained by local representatives by knowing needs and representing them to the state and elected *arrondissement* officials' quest for legitimacy in the eyes of the central city executive.

The changes in position and action repertoires effected in *arrondissements* after 1995 clearly reflect a dynamic of institutional construction, materialised through the activity of creating bodies and permanent official links that enabled the emerging institution to occupy, equip, and organise its territory. The legislatively created infra-municipal echelon, by becoming in its turn a creator of institutions, gradually acceded to the rank of full-fledged institution. The work of constructing the territory and proximity accomplished by *arrondissement* mayors, who are also political leaders anxious

[13] İt should be noted that official *arrondissement* council recommendations increasingly result from informal pre-consultation activity and exchanges between *arrondissement* representatives and the relevant central city departments and deputies.
[14] On the neighborhood council experiment in the 20th *arrondissement* of Paris, see Houk 1996; Blondiaux and Levêque 1998 and the 1997, 1998, and 1999 Rapports de l'Observatoire de la Démocratie Locale du 20ème.

to have a substantial existence in the municipal opposition camp, thus partakes of a process of institutionalisation in Selznick's sense: proximity 'acts'; it is mobilised as a value giving specific meaning to infra-municipal action and the infra-municipal institution.[15] The actions and processes just considered called into question the idea that *arrondissements* existed and functioned undifferentiated from the central city government apparatus – precisely the assumption on the basis of which the first *arrondissement* mayors had developed their offices and activities. The power struggle with city hall initiated and maintained by the new *arrondissement* mayors should be measured in terms of that change.

Differentiation Through a Power Struggle with Central City Government

Because they were in the opposition camp, and in the name of the legitimacy conferred on them by their political majority in the locality, the leftist mayors elected in Paris and Lyon in 1995 were quick to demand a greater role in managing district affairs. There had been occasional 'rebellious' moves against the higher echelon during the 1983-89 office term in particular; Pierre Bas, mayor of the 6th *arrondissement* from 1983 to 1995, had been especially active, refusing to approve the level of funding allotted by the central city government or projects that did not serve the interests of his *arrondissement*. But the actions or reactions of other district mayors involved much more discreet 'corridor' negotiations, in the framework of bilateral relations obtaining between individual mayors and Chirac's cabinet.[16] A modus vivendi had been quickly established: central and local echelons exchanged points of view before centrally designed projects were sent to *arrondissement* councils so that those deliberations would not become a matter of settling scores with city hall. Starting in 1995, however, media attention to conflicts between the two levels made the power struggle more visible, particularly in Paris.

The power struggle with central city government was constructed along three major lines: district mayors' move to 'better apply' the law; *arrondissement* resistance to projects sent down from the centre; district officials' participation as territorial representatives in new territory-focused public policy arrangements and procedures.

The fight for a different interpretation of the Loi PML was waged most fully in Paris. A number of newly elected mayors, led by Michel Charzat, Roger Madec, and Georges Sarre, mayors of the 20th, 19th, and 11th *arrondissements*, brought the issue to court, accusing the central administration of unequal treatment for majority and minority *arrondissements* and attacking the way the text was interpreted, going so far as to claim that some provisions were not being applied at all. The way the law was interpreted at the time was consistent with the uniform political situation produced by the 1983 and 1989 city elections and Chirac's consequent subjugation of *arrondissement* mayors, orchestrated with their complicity. District councils and the central city council had agreed on a number of special exceptions to the law with regard to

---

[15] Selznick defines institutionalisation as a process - 'to institutionalise is to infuse with value' (p.17) - in which the leader plays a major role: 'the institutional leader is primarily an expert in the promotion and protection of values' (p.28).

[16] From 1983 to 1995, it was Jean Tiberi, mayor of the 5th *arrondissement* and a close ally of Chirac, who was in charge of relations with *arrondissement* mayors.

transfer from city to district of responsibility for certain facilities and what proportion of the meagre amount of available public housing was to be allotted by *arrondissement* authorities.[17] The 'choice' by *arrondissement* mayors not to exercise certain functions that the law vested in them was of course in part strategic. With regard to housing, for example, since the number of units available was much lower than the ever-increasing demand, it was actually in *arrondissement* representatives' political interest not to have to manage this procedure, especially when it came to turning down requests from hopeful families.

Paris opposition mayors now demanded that three of the law's provisions be applied more rigorously. Specifically, they demanded a clear inventory of what facilities were to be managed by what level (this implied that facilities not then under *arrondissement* management due to local-centre majority agreement but authorised to be so under the law, particularly socio-cultural and sports facilities, would be transferred to them); a clear indication of proportion of public housing to be allotted by district mayors (this implied transparency about number of units available and verification that they were indeed allotted by *arrondissement* mayors), and a clear basis for calculating level of funding allotted to *arrondissements* (this implied improving accounts of each *arrondissement's* demographic and socio-occupational characteristics). The administrative court ruled across the board in favour of applying the law in ways more favourable to *arrondissements*. The main point, however, was that though official *arrondissement* jurisdiction and responsibilities were not modified by the court ruling, opposition leaders could capitalise on these legal victories in terms of visibility and political image.

The proposals for reforming the law that began to come in 1997 were consistent with this strategy. With their stated objective of increasing infra-municipal echelon responsibilities in all matters concerning daily life, reform projects submitted by MPs, who, it should be remembered, were locally elected, attracted major media attention. Indeed, this was their real purpose, as is clearly illustrated by the in-pouring of proposals from across the political spectrum as the 2001 elections approached: these projects had not won majority support in the various political apparatuses and were hardly likely to come to fruition.[18]

Another type of differentiation through power relations was leftist mayors' resistance to projects sent down by the city majority. That resistance intensified after 1995, particularly in Paris. The newly elected mayors at first worked to test the limits of the legal provisions (as their predecessors had done, but less discreetly), particularly the stipulation that city governments must consult *arrondissement* councils on all matters affecting the *arrondissement*. Since the city hall is not required to follow *arrondissement* council recommendations, voting against a project proves a limited

---

[17] Article L2511-20 of the CGCT specifies that half of the public housing that the mayor of Paris makes available to a given *arrondissement* is to be allotted by the *arrondissement* mayor. For studies of early implementation of the Loi PML in Paris, see Souchon-Zahn 1986; Knapp 1987 and Haddab 1988.

[18] The 'démocratie de proximité' law of 28 February 2002, only slightly modified the Loi PML and left *arrondissement* status and official responsibilities unchanged, whereas reform projects submitted by officials on the left proposed among other things to make *arrondissements* legal entities.

means of action. But it does allow the opposition to publicly express its disagreement and above all to show that city hall is not much interested in what local majorities thought (this was done first at *arrondissement* council meetings, then in communiqués to *arrondissement* residents and associations). Moves were made to refuse to vote or postpone voting in protest against how little time the central government allotted for getting its projects on *arrondissement* agendas and how little information it made available to local representatives who nonetheless had to produce a recommendation. In its first years in office, the 20th *arrondissement* mayoral team refused several times to vote on projects involving subsidies to local associations. In doing so, they succeeded in collecting new information on the nature and activity of city hall subsidised associations. But overall, the strategy of obstructing procedure was used only occasionally, and less and less frequently over time. Today local officials favour a vote of approval combined with a statement of *arrondissement* majority's position or specific requests for changes in projects (reduced number of storeys for a real estate development, change in ZAC public housing plans, planning for public-use facilities or grounds, request that city government purchase an abandoned industrial lot, etc.).

Meanwhile, leftist mayors, in search of ways to compensate for their weakness in the face of city hall power, worked to gain the support of *arrondissement* social forces for official council positions. Postulating that the city government majority could not ignore local demands or preferences when expressed by a collective of citizen voters, opposition officials tried to develop local mobilisation around controversial projects, either by pledging support and advocating for them to the central government or working 'underground' to get the local society (individuals or associations) to put pressure on the city government (petitions, parent delegations sent to city hall to protest against the closing of a school class, for example). In this respect, neighbourhood councils came to be instruments in the service of the power relations that *arrondissement* officials have engaged and seek to maintain with the central executive: neighbourhood council recommendations help legitimate *arrondissement mairie* opposition to projects its citizens 'do not want'. Neighbourhood councils, spurred on by *arrondissement* authorities, can also organise groups to lobby or pressure the central authorities.

The last action repertoire involving a dynamic for differentiating the infra-municipal echelon and used by district representatives increasingly after 1995 was to become active in specific public policy arenas. To have a presence in the urban political landscape, local representatives of every political stripe began seizing the opportunity offered by the new state policy of 'territorialising' public policy. Since the early 1990s in France, public action has been based on policy designed for clearly delimited territories and on 'contract' arrangements that bring together several partners (the state, region, *département*, city, and so forth). The overarching urban policy known as *la politique de la ville* is emblematic of this change and its instruments have been applied to significant parts of Paris, Marseille, and Lyon neighbourhoods since the mid-80s.[19]

---

[19] The most recent *politique de la ville* instruments, entitled Développement Social Urbain (DSU) and Grand Projet de Ville (GPV), define the perimeters within which specific cross-cutting actions, aimed at spurring development or improving living conditions and jointly financed by all different levels of government, may be conducted.

Just as city government deputies were doing, district mayors and their deputies began working to gain clout in policy steering committees as legitimate representatives of the neighbourhoods concerned. Given their presence on the territory, familiarity with the daily running of associations that were now official policy partners, and knowledge of local realities, they have not had much trouble being admitted to work with Développement Social Urbain steering committees, for example. More recently, the national policy of establishing contracts in the areas of security and crime prevention (Contrats Locaux de Sécurité or CLS) has been implemented in Paris, Marseille and Lyon, in association with district representatives, on the basis of district C.L.S., enabling them to become institutionally involved in the drafting and follow-up of contracts for the security policy of their territory.

But *arrondissement* officials desirous of being seen by the central city government, services in their *arrondissement*, and the population at large as references and forces to be dealt with, encountered/still encounter/have encountered strong resistance from the municipal apparatus. Though the city authorities now formally recognise that *arrondissement* mayors and deputies have a place in such bodies as policy steering committees, they fully intend to maintain control over what transpired in them. The central city government of Lyon (politically on the right) set up its 'Mission 8ème' just next to the 8th *arrondissement* hall where the left had come to power in 1995. This municipal bureau, the 'Mission 8ème', is charged with following urban projects underway in the *arrondissement*; its director was positioned to be the main interlocutor for various urban planning actors operating on the ground, including those in charge of *politique de la ville* actions. In this way, city hall kept up a kind of competition between central and local echelons, keeping central government's local information channels open and preventing *arrondissement* representatives from acting as intermediaries between the municipality and actors on the ground. Lastly, their presence in the *arrondissement* enabled them to monitor the actions of *arrondissement* representatives.

The various action repertoires I have evoked here, put in place after 1995 by opposition *arrondissement* mayors though they have gradually come to be used by members of the municipal majority too, fuelled the dynamic for differentiating the infra-municipal echelon from that of central city government. Thus constructing the institution by means of the territory and the territory by means of the institution worked to affirm the infra-municipal echelon as a full-fledged entity, though not one that can be qualified as autonomous since city hall has the financial resources and final decision-making power for *arrondissements*. Nevertheless, the affirmation of the *arrondissement* level has led to the development of a kind of competition between infra-municipal and central echelons, not so much around decision-making power as presence on the ground and direct access to citizens. The *arrondissement* authority appropriated the concept of proximity and used it as a means of constructing its own identity. Clearly what was at issue in this process was differentiating the two territorial echelons.

Conclusion

In analysing how the Loi PML was implemented, I have looked at the emergence of a new institution, the forms and action repertoires by means of which it was gradually constructed, and the dynamics that structured its position. This approach must take into account the fact that an institution is part of, and sometimes manages to gain a degree of power and influence in, a pre-existing environment that not only shapes it but reacts to it and adjusts in response to its existence and behaviour.

In the case of *arrondissements*, there are clear signs that the infra-municipal echelon is now recognised by the environment in which it has affirmed itself and that this recognition is a reaction to recent changes in its position. Most remarkably, the central level has adjusted its behaviour and functioning to take into account and recognise the institutional existence of *arrondissement* authorities.

Arrondissement mayors in the three cities now participate at an earlier stage and more systematically than before in the development of municipal plans involving their territory. They sit, for example, on the occasional and ongoing theme-based commissions set up by city hall for urban planning projects (housing, tramway lines, etc.) or sector policy (security, urban transport, etc). Some formal arrangements have been made to improve communication between central and local echelons and the way they work together, particularly in Lyon, where Raymond Barre charged his first deputy to reflect on the question immediately after taking office. The resulting 'Rapport Philip' recommended a new procedure for city council project deliberation that would bring in the relevant *arrondissement* mayor and his/her team early in the process, before official presentation of projects to the *arrondissement* council. Another adjustment resulting from the report is to allow *arrondissement* deputies (who are generally mere district councillors and as such have no direct access to the city executive) to participate, though not to vote, in city commissions relevant to their area of concern. Here it is very clear that *arrondissement* mayors' self-assertion after 1995 has opened up existing city government systems, making them take fuller account of certain local realities and resident demands.

The administrative municipal services in charge of *arrondissement* authorities in Lyon (Direction des Mairies d'*Arrondissement*) and Paris (Bureau des Mairies d'*Arrondissement*, part of the Direction de la Vie Locale et Régionale) are gradually taking on substance. Focused at first on *arrondissement* administrative functioning, and consultation with *arrondissement* councils, these services are gradually becoming intermediaries between the central and local political spheres, at times playing the role not only of informants but also buffers for sensitive or controversial issues.

As explained above, there are territorial aspects to certain contractual policies (local security contracts, *politique de la ville*, urban transport plans, for example), and these make the *arrondissement* and its representatives legitimate interlocutors. They also attest to recognition on the part of institutional actors other than the central city government.

Lastly, institutional recognition of the infra-municipal echelon is materialising – and being fuelled by – the emergence of the *arrondissements* as full-fledged political stages. Though it would be excessive to say they have become autonomous from the municipal political scene, recent behaviour does attest to the fact that the *arrondissement* is more than a mere stop along the way in political careers aimed at higher offices

(Houk 2001).   In the 2001 municipal elections in Marseille, four incumbent *arrondissement* mayors headed the city majority party's ballot in their sector, and then took up their sector offices again.   In Paris, incumbent *arrondissement* mayors from the officially unified right majority presented themselves for the first time on dissident ballots.   15th *arrondissement* mayor René Galy-Dejean refused to yield his place at the top of the ballot to Edouard Balladur, former presidential candidate and MP, who had been designated once again by the RPR and UDF political apparatuses to head the ballot for that *arrondissement*.   Galy-Dejean chose to play his own political card, and won.   In Lyon, Gérard Collomb, socialist mayor of the 9th *arrondissement* from 1995 to 2001, was voted to lead the city council.   Though the effects of the local political situation i.e. the serious dissent and division within the forces of the right in Lyon and Paris, should not be underestimated, these developments are also, in my opinion, to be accounted for by the infra-municipal echelon's new-found strength in the city landscape.

These phenomena all attest to an institutionalisation process which, though not complete, is now sufficiently advanced to justify the claim that *arrondissement* authorities are more than a mere component of today's Paris, Marseille, and Lyon city governments.   Moreover, they show the issues involved in managing proximity, particularly in the framework of large authorities like the three cities studied here, but also for new forms of intercommunal cooperation, such as *communautés d'agglomération* [50,000 inhabitants, including a city of at least 15,000]   and *communautés urbaines* [500,000 inhabitants, including a city of at least 50,000].   The question arises of how the different levels of political representation and official territorial jurisdiction fit together.   There are regular public debates on whether there should be intercommunal assemblies elected by universal suffrage or whether official proximity-managing responsibilities should be attributed to *arrondissement* authorities.   Such proposals, though just as regularly put aside, are always an occasion for taking up the issue of the 'necessary' coordination of institutional structures that have been established gradually, one on top of the other.   Broadly speaking, the debate corresponds to a renewal of concern about local democracy and citizen participation practices.   In the three cities studied, the firming up of the infra-municipal echelon has undeniably led city governments to reconsider their own definition of proximity and the ways they relate to urban society.   In the case of Paris, the post-1995 initiatives of leftist mayors in matters of participatory democracy have encouraged, and perhaps in certain cases provoked, an opening up of the municipal apparatus to new consultation and dialogue practices, namely in the area of urban planning.   This development fits, of course, with broader changes in political practices and public action, all involving greater citizen participation, but it has been much facilitated by the way opposition mayors have positioned themselves.   In Marseille and Lyon, where central city governments long relied on a fabric of neighbourhood committees, the new activities of *arrondissement* representatives are part of a decided renewal of both the frames and types of actors considered legitimate participants in local dialogue in urban communities.   In this respect, there is no doubt that *arrondissement* officials in the three cities have a role to play in restoring participation of citizens as members of a political community rather than mere users of urban services, especially since what is at stake in strengthening the tie between municipal government entities and civil society is the renewal of political legitimacy itself.

References

Bagnasco, A. and P. Le Galès (ed) (1997) *Villes en Europe*. Paris, La Découverte.

Baraize, F. and E. Négrier (ed) (2001) *L'invention politique de l'agglomération*. Paris, L'Harmattan.

Blondiaux L. and S. Levêque (1998) La politique locale à l'épreuve de la démocratie : les formes paradoxales de la démocratie participative dans le 20ème *arrondissement*. In Neveu C. (ed) *Citoyenneté et territoire*. Paris, L'Harmattan.

Blondiaux, L. (1999) Représenter, délibérer ou gouverner? les assises politiques fragiles de la démocratie participative de quartier. In (coll.) *La démocratie locale, Représentation, participation et espace public*. Puf, pp.367-404.

Borraz, O. (1996) *Gouverner une ville. Besançon 1959-1989*. Presses Universitaires de Rennes.

Borraz, O. (1995) Politique, société et administration : les adjoints au maire à Besançon, *Sociologie du Travail*, Vol.37, 2/95, pp.221-248.

Caillosse, J. (ed) (1994) *Intercommunalités*. Presses Universitaires de Rennes.

Di Maggio, P.J. and W.W. Powell (1983) The Iron Cage Revisited : Institutional Isomorphism and Collective Rationality in Organisational Fields, *American Sociological Review*, No.47, pp.147-160.

Garraud, P. (1989) *Profession : Homme politique – La carrière politique des maires urbains*. Paris, L'Harmattan.

Haddab, K. (1988) L'application de la loi P.M.L. à Paris ou le centralisme à l'échelon d'arrondissement, *Annuaire des Collectivités Locales*, pp.67-84.

Houk, M. (1996) *Les conseils de quartier du 20ème, une expérience de démocratie locale*. Mémoire de DEA de Sociologie, sous la direction de O. Borraz, I.E.P. de Paris.

Houk, M. (2001) Vers une décentralisation municipale à Paris?, *Esprit*, No.6, June 2001, pp.193-200.

Joliveau, T. (1987) *Associations d'habitants et urbanisation – L'exemple lyonnais 1880-1983*. Mémoire et Documents de géographie, Paris, Editions du CNRS.

Knapp, A. (1987) Le système politico-administratif local parisien : 1977-1987, *Annuaire des Collectivités Locales*, pp.65-90.

Le Galès, P. (2002) *European Cities*. Oxford, Oxford University Press.

Lorrain, D. (1981) *A quoi servent les mairies ? La gestion municipale ou la régulation au quotidien*. Paris, Rapport de recherche pour la Fondation des villes.

Lorrain, D. (1991) De l'administration républicaine au gouvernement urbain, *Sociologie du Travail*, 4/91.

Mattina, C. (2001) Des médiateurs locaux : les présidents des C.I.Q. autour de la rue de la République. In A. Donzel (ed) Métropolisation, gouvernance et citoyenneté dans la région urbaine marseillaise, Paris, Maisonneuve et Larose, pp.269-291.

Meyer, J.W. and B. Rowan (1977) Institutionalised Organisations : Formal Structure as Myth and Ceremony, *American Journal of Sociology*, No.83, pp.340-363.

Nay, O. (1997) *La région, une institution*. Paris, L'Harmattan.

Négrier, E. and B. Jouve (ed) (1998) *Que gouvernent les régions d'Europe?* Paris, L'Harmattan.

Olive, M. (1997) L'organisation administrative de Paris, Lyon et Marseille : mise en œuvre d'une réforme institutionnelle. Le cas de Marseille. In D. Gaxie (ed) *Luttes d'institutions – Enjeux et contradictions de l'administration territoriale*. Paris, L'Harmattan, pp.195-232.

Pôle Sud (2000) numéro sous la direction de J. Joana, *Qui Gouverne les villes?* No.13.

Rapports de l'Observatoire de la Démocratie Locale du 20ème, 1997, 1998 et 1999.

Sellers, J.M. (2002) *Governing From Below, Urban Regions and the Global Economy*. Cambridge University Press.

Selznick, P. (1984) *Leadership in Administration*. Berkeley, University of California Press.

Souchon-Zahn, M-F. (1986) L'administration de la Ville de Paris depuis 1983, *Revue Française d'Administration Publique*, No.40, pp.677-707.

Chapter 14

Metropolitan Democracies: From Great Transformation to Grand Illusion?[1]

Bernard Jouve

Introduction

The 20th century rested on a set of basic principles relating to the political organisation of modern Western societies. Indeed, it was for a long time inconceivable to imagine politics outside the context of the nation-state, with its institutions and its territory controlled through a set of borders and normative instruments. Both the social sciences and the political realm used this three-layered concept of state, society and territory as a unit for analysis and action. There could be no society other than a national one, regulated by the state, within a defined territory. The dominant system of legitimacy in Western states accorded primacy to representative democracy over any other alternative form, such as participatory or deliberative democracy. It was because national institutions (especially legislative and/or executive bodies and administrations expected to reflect, without any interference, the directions and choices established by the political realm), benefiting from the legitimacy conferred through election by universal suffrage, could develop and implement public policies that responded to the 'needs' of their constituents in the name of public interest, even when this involved using coercion. This political model appears to have outlived itself and has, at the very least, been called into question.

During the mid-1970s, the first signs that this form of state-centred political organisation had run out of steam were identified. In a report that became well known, Crozier, Huntington and Watanuki referred to a 'crisis of democracy' in the West, mainly expressed through the inability of nation-states to meet all the social demands that were being made of them. This 'overload' of the state apparatus led to an incapacity to act and, therefore, to its functional legitimacy being called into question (Crozier et al. 1975). In the 1980s, in the context of the 'conservative

---

[1] I would like to thank F. Bardet, P. Booth, J.-A. Boudreau, O. Borraz, A.-G. Gagnon, P. Hamel, P. Le Galès, D. Latouche, C. Lefèvre and E. Négrier, H. Savitch and M. Zepf, as well as the members of the *Groupe de Recherche Interdisciplinaire sur les Mouvements Sociaux (Montréal)* (interdisciplinary research group on social movements) for their comments, highly pertinent as always, on the first draft of this text.

revolution' in the United States and Great Britain, this 'crisis' theme disappeared from the political agenda for a time. Recourse to a free-market formula, based on deregulation and privatisation, partly explains this development.

It was only in the mid-1990s that mention was again made of this 'crisis of modern democracies', brought about by globalisation, the reconstruction of nation-states, and through other major sociological changes. The terminology also changed. The 'crisis of modern democracies' became the 'crisis of governability' requiring new tools of governance (Kooiman 1993). This semantic change allowed emphasis to be laid on the fact that the 'crisis' was not expressed solely in terms of an 'overload' of the state apparatus, but more fundamentally in the context of a two-sided challenge, on the one hand, to the very conditions behind public policy-making and, on the other hand, to the legitimacy of the public authority itself. In fact, the dynamics behind this double challenge began to appear in the 1960s. In particular, we might mention:

- The calling into question of a form of political practice based on domination (Mayer 2000);
- The calling into question of the primacy of political parties as authorities representing the collective preferences of individuals and of partisan affiliations and electoral loyalties (Pharr and Putnam 2000);
- Criticism of a mode of collective representation of preferences based on legal-rational legitimacy, monopolised by the nation-state and its administrations (Habermas 1997);
- A case being made against a liberal conception of the modern state, in theory open to the range of claims coming from civil society. The most venomous criticism of this concept came from Marxist authors who linked the crisis of the nation-state to changes in capitalism (Brunhoff and Poulantzas 1976);
- A loss of confidence in politics itself, as regards its ability to deal with all the problems of modern society, and the emergence of a civil society which is making increasing demands in terms of the organisation of power (Keane 1998);
- The fragmentation of decision-making systems following changes to the internal structure of states due to decentralizing reforms and federalist dynamics (Loughlin 2001);
- The challenge to a decision-making model based on political representation (and therefore on the central role of elected political officials) and on the primacy of academic expertise used by administrations hiding behind their monopolisation of technical skills (Callon et al. 2001);
- The emergence of new issues (the environment, exclusion, integration) that can no longer be dealt with through sector-based policies, requiring instead an integrated approach and the search for synergies between institutions with different logics of action, cultures and temporalities (Duran and Thoenig 1996);
- The consolidation of new territories of collective action, in particular metropolitan cities within which social movements had, since the 1970s, made a case against political integration 'from the top' (Hamel et al. 2000a);

- Lastly, the redefinition of citizenship, in its liberal and universalist expression, by social groups demanding a community-based treatment which they feel will allow them to disregard policies, which, in the guise of political liberalism, hinge on the discrimination of groups that are disadvantaged on the basis of gender, language, ethnic origin, religious practice, or sexual orientation (Beiner and Norman 2001).

It is in this general context in which a state-centred model of politics based on representative democracy, questionable academic expertise, and a universalist conception of citizenship was being questioned, that the themes of local democracy and political proximity slowly became established in both the academic and political spheres. The local level, understood as a two-dimensional physical and political space, is becoming once again the new reference territory for politics based on which it is deemed possible to rethink and act on the crisis of governability in modern societies, and to solve the set of problems referred to above. The scope of this dynamic varies with each national political context. In the United States it is a major feature of the construction of the nation-state; 'local' is synonymous with political proximity, efficient democracy and, above all, institutions that offer an alternative to the federal government. In Europe, on the other hand, this process is more recent. In both cases, the local level is seen as the level at which all social, economic and political contradictions can be overcome. Having become the stuff of myth (and at times the object of a cult-like reverence) it has the status of the privileged level of authority, based on which it is possible to ensure the transition from the 'first modernity', which established the central role of the state in the political and economic organisation of societies, to a 'second modernity' which is in sync with globalisation and offers a solution to the crisis of institutions (Giddens 1990; Beck 1992; Dubet 2002).

More specifically, the chapters assembled here focused on metropolitan cities as new sub-national political spaces. This is not to deny the potential importance of other territories, such as regions, or even the supra-national level. In their chapter, Duchastel and Canet fully demonstrated the importance of seeing the transformation of modern democracies in a local-to-global continuum. The decision to focus on the metropolitan level in these three countries can mainly be explained by the demographic significance that the urban reality has acquired on a global scale since the post-war period. According to the United Nations, the rate of global urbanisation rose from 29.8 per cent in 1950 to 47.2 per cent in 2000. This rate is currently 77.4 per cent in North America and 73.4 per cent in Europe.[2] The continuation of the urban transition in developing countries and the new forms that urbanisation is taking on in developed countries suggest that, in the medium term, the problems of political regulation faced by public authorities will merge with those of the metropolitan city and/or will take place in the following arenas: exclusion, socio-spatial segregation, immigration, public safety, the environment, health, education, and sustainable development.

This subject, however, is often difficult to fully grasp. What distinguishes a metropolis from a city? What makes it a unit? Is the legal and institutional

---

[2] http://unstats.un.org/unsd/demographic/default.htm Consulted on April 8, 2004.

interpretation pertinent? Not necessarily, because there is almost always a discrepancy between the so-called metropolitan institutions and the functional territories. The Greater London Authority, the Municipality of Paris, and Greater Lyon, to cite a few examples, are institutions that do not match the populated spaces that structure London, Paris or Lyon. Greater London, in reality, represents the southeastern quarter of Great Britain and not merely the space covered by the 32 boroughs and the City that compose the Greater London Authority. The Parisian metropolis stretches across the whole territory of the Île-de-France region. The functional territory of Lyon is spread over, not only the 55 *communes*, represented by the *Conseil de la Communauté Urbaine*, but also over 678 municipalities covering a space of 8280 km². The discrepancy between the geography of the population flow and the metropolitan institutions is enormous. Would the demographic approach be more effective? The academic discussion concerning the threshold at which a city becomes a metropolis has not yet been concluded, and most likely never will be. We could opt, as Le Galès does, for a historical approach and stress the fact that, in Europe, medium-sized cities are at the origins of the development of the modern state (Le Galès 2002), while the link between the large metropolitan cities such as Paris or London and the construction of nation-states is less well-established. This approach, however, is not well suited to the situation of 'new countries' such as Canada.

In order to define the subject of our research, an economic and socio-political approach seemed more productive. Indeed, the metropolitan cities constitute the territories that are most directly affected by the economic transformations that accompany globalisation. It is within them that the process of 'creative destruction' (Schumpeter 1942) can be seen with the greatest force, calling into question the balance of power and the pre-existing socio-economic order, and bringing about the creation of opportunities for some social groups while, at the same time, creating new social inequalities. Moreover, it is within these territories, with their very hazy geographical and institutional boundaries, that the redefinition of politics and the transformation of the mechanisms of integration and differentiation are taking place. The 'big city' was the favoured space of sociological modernity (Simmel 1989). The 'second modernity' of Beck (1992) is currently found in the metropolitan cities. They represent the territories in which the transformation of the social contract that ties the political realm to civil society in modern states is taking place. The basis of the political order in Western democracies has for a long time been based on two major principles: the central role of elected representatives, who benefit from the legitimacy that comes from being elected; and a universalist approach to citizenship, as organised by the state. These two principles are currently being called into question through a two-edged demand coming from civil society for the opening up of decision-making systems and the recognition of the difference between various social groups, based, for example, on language, ethnic origin, race or religion. The goal of this book was precisely to analyse this transformation, which justifies the decision not to limit the empirical subject to one type of metropolis, one institutional form or one territorial level in particular. The process underway is developing within metropolitan institutions, within municipalities and, on a sub-municipal scale, within various districts in the metropolitan cities. From this point of view, this book is based on the dual hypothesis that metropolitan cities are not just sub-national territories, but that the socio-political changes taking place within them affect the very foundations of the political order, these being the primacy of elected political representatives and a universalist approach to citizenship. Regarding the dynamics that are developing in other subnational spaces, such as regions or rural territories, there is an inherent difference in the phenomena, and not simply a difference of degree in the extent of the processes taking place there. Do the various chapters of this book generally confirm or partially invalidate this dual hypothesis? This is the question addressed by this conclusion.

Economic and socio-political dynamics can also take place on the fringe, or even outside of metropolitan institutions, thus making it more difficult to address them, given that there are no clearly identified institutions responsible for them. This lack of co-ordination between institutional territories and the spaces where issues arise is, however, not synonymous with an inability to govern. For several years now, the literature on urban governance has highlighted the possibilities for 'governing differently', that is to say other than through democratically elected institutions, and more specifically from within network-based institutional arrangements, through partnerships which call upon different types of actors to intervene in various territories (Le Galès 1998; Leresche 2001). It is not so much the question of political management 'on the fringe' of institutions that poses a problem; more the issue of citizen participation in these new public policy arenas, which may well be effective but that must face up to a deficit of democratic legitimacy.

Therefore, the 'metropolis' – or 'urban region' or 'city-region', as it is also referred to (Pumain 1993) cannot be grasped based on a predefined set of geographical, institutional, or demographic variables. This is what mainly distinguishes it from the 'urban' level. The methodological construction of the subject is difficult and requires appropriate tools (Moriconi-Ebrard 1993; Jaccoud et al. 1996). 'Metropolitan cities' refers here to cities that have a particularly significant connection to globalisation and which are, at the same time, spaces within which the political order is being transformed through the calling into question of the central role played by elected officials, the desire for decision-making modes based on political participation, and the challenge to a universalist-type system of citizenship. Our subject, then, is not confined to 'global cities' or 'global city-regions' (Sassen 1996; Scott 2001) which refer to the urban communities at the top of the global hierarchy in terms of population and centralised leadership roles. The argument defended here is that this double dynamic of the integration of metropolitan cities into the world system and the questioning of the basis of the political order also concerns seemingly less 'central' spaces. This methodological decision explains why this book is made up of chapters dealing with very different spaces and institutions. It is the processes of transformation of the relationship between civil society and politics that define metropolitan cities, rather than institutions or any particular territorial level or demographic threshold.

It is important not to approach democracy within these spaces from a rosy or romantic perspective that would deny the existence of a structuring of modern societies that is still and will always be characterised by inequalities in the allocation of resources, powerful mechanisms of exclusion, and the domination of some groups by others. In fact, metropolitan cities are the territories in which new forms of poverty are springing up with the greatest force (Musterd and Ostendorf 1998; Schnapper 2001; Paugam 2002). The current mechanisms of socio-economic exclusion are no longer confined solely to the social division of labour within production relations. In fact, they compound a more extensive problem concerning the very existence of social ties between social groups that are included in, and those that are excluded from, the production system. More fundamentally, the 'new urban question' (Donzelot 1999) involves the combination of exclusion from the production system and disaffiliation from the political system. These two processes have the most negative effects when they involve specific racial groups or immigrants (Verba et al. 1995; Body-Gendrot and Martiniello 2000). This is where the question concerning the practice of democracy within metropolitan cities becomes so important. Do democratic practices within these territories allow for a political re-affiliation of the most disadvantaged groups? Do they sustain a transformation in the relationship to politics that, then, ceases to be organised on the principle of representation? Is it possible to identify the metropolitan level, and no longer the municipal level, as the space in which this transition is taking place? To what extent are decision-making systems opening up to new social, community or economic actors? In the context of the pluralisation of 'urban regimes', what is the significance of the re-organisation that is taking place in the relations between metropolitan actors and the state? Are we witnessing the emergence of a metropolitan citizenship and if so, is it based on a universalist/republican model, or is it community-based? This book aimed to address these questions by examining the dynamics of convergence and divergence between British, Canadian and French metropolitan cities.

Participatory and Deliberative Democracy as a Political Remedy and Management Principle

It would be incorrect to maintain that the 'political crisis' is affecting politics as a whole, understood as a system integrating ideological elements as well as institutions, specialised actors, elected representatives, and citizens (Norris 1999). Within Western societies, there is in fact a general consensus as to the values that liberal democracy represents. However, the institutions and actors that are supposed to put these values into action should take note because (and to varying degrees depending on the state in question) it is political parties and elected representatives that are eliciting the most discontent (Pharr and Putnam 2000; Putnam 2002). It is in this context that recourse to participatory democracy in cities is largely seen as the solution to the problem of mistrust regarding politics. In essence, this type of democracy attempts to form the mould from which the relationship between the political realm and civil society can be completely recast. Essentially, the local level seems to be turning its back on representative democracy, preferring participatory and deliberative democracy. While locally organised participative democracy in different states and metropolitan areas serves as a vehicle for very different kinds of representation and discourse (Boudreau 2003), inspired by the work of De Tocqueville (1842), ailing modern democracies are increasingly calling upon participative democracy to help reconstruct a political order

Certainly local democracy, based on the principle of participation and no longer solely on political representation, is supposed to develop a sense of belonging to a community, commitment, generosity, a sense of morality, interest in public affairs, and the desire to transcend individual interests. Indeed, these are all virtues and examples of 'civic competence' that are said to distinguish representative democracy from participatory democracy (Elkin and Soltan 1999). Furthermore, at a time when the role of the state in the regulation of modern societies is being increasingly challenged, the significance of J.S. Mill's work on local democracy in the United States is being rediscovered. In his book, *On Liberty*, J.S. Mill expressed the opinion that the state should not interfere in the decisions of citizens, who are in the best position to make choices that match their needs and preferences (Mill 1859). It goes without saying, however, that this enchanted vision of local democracy does not hold up very well in, even in the small towns of New England (Mansbridge 1980; Bryan 1999). In fact, several authors in the United States have identified a civic and political disengagement, considered to be a serious cause for concern from the point of view of the general functioning of society (Skocpol and Fiorina 1999; Putnam 2000). Many Western democracies are trying to prevent precisely this kind of decline in the relationship to politics.

Thus, many observers consider the local level as the space in which there can be a renewal in politics, and in which an effective political system can be rebuilt by mobilising civil society around collective problems before it is too late and democracies begin to suffer from the 'Bowling Syndrome.'[3] With the appointment in France of Raffarin as prime minister in 2002, proximity became the central theme on which the state is to be radically reformed. In Great Britain, the rise to power of New Labour in 1997 was also mainly backed by a discourse that focused on the need to redefine the place of citizens in public policy making, while remaining mistrustful of local elected representatives. In a more general sense, as demonstrated in the chapter by Dabinett, the procedures used to develop and implement urban policies in Great Britain have clearly been based on the principle of proximity and the participation of residents. In Quebec, the election of the Liberal Party in 2003 was also based on a platform that criticised bureaucracy and the development of a government apparatus judged to be excessive and that promoted the need to carry out a 're-engineering' of the Quebec government.

The theme of political participation, deliberation and proximity has thus become a 'must,' particularly in urban politics (Blondiaux and Sintomer 2002). Aside from the context identified by A. de Tocqueville and J.S. Mill a century and a half ago, two major dynamics explain the ongoing pertinence of this theme:

---

[3] Reference is made here both to the writings of Putnam and the cinematographic work of Michael Moore. In his work on the erosion of civic engagement within civil society in the United States, Putnam uses bowling as a particularly relevant analytical prism. Long considered a veritable institution ensuring the construction of 'social capital', bowling, according to surveys by the author, has lost its function of social integration because the sport is no longer played in groups, but rather by individuals playing alone. Moore, in his film 'Bowling for Columbine', refers to the mass killings perpetrated in a school by two teenagers who, several hours before committing the unpardonable deed and then taking their own lives, had nonchalantly gone bowling together. This is what is referred to here as the 'Bowling Syndrome', which characterises a society where social ties are breaking down, community relations based on mechanisms of organic solidarity are crumbling, and political participation is disintegrating, mainly to the detriment of disadvantaged groups.

- Participatory democracy as instrumental in loosening categories and liberating disadvantaged social groups. Aside from the success, (in particular the recent media success, of Porto Alegre and its participative budget, Gret and Sintomer 2002), it was again in the United States, as of the 1960s, that the analytical basis was laid, establishing a link between local participatory democracy and the balance of power between urban social groups (Kaufman 1960; Pateman 1970; Bachrach and Baratz 1975). Fuelling the New Left Movement in the United States, this analysis was the basis of some public action programmes such as the Economic Opportunity Act of 1964, the Model Cities Program of 1966, and the Urban Renewal Project of 1968, run by the federal government and aimed at the black community living in the inner cities. Through measures targeting this particular community under the banner of 'neighbourhood government', the goal was to foster civic engagement within underprivileged neighbourhoods and to allow for the emergence, through training and political socialisation programmes made possible by participation, of a political elite developing out of disadvantaged groups, while controlling the selection process to prevent this elite from adopting positions judged to be too radical or even extremist (Hallman 1974; Schoenberg and Rosenbaum 1982). First applied to the inner cities in the United States, this formula has since been exported to several countries, in particular Great Britain and France (see the chapters by Dabinett and Chabanet);

- The importance attached to deliberative activity as the basis of legitimacy in modern democracies, within which the definition of the common good is no longer vested in a single institution – the state. It was initially based on the public space approach developed by Habermas that a debate took shape, emphasising the fact that in representative democracies the legitimacy of elected representatives and administrative officials now rests on the production of discourse around the definition of the common good. These discussions, carried out in the context of interactions with 'ordinary' citizens, allow for a plurality of viewpoints to be heard and for prevailing values to be called into question (Habermas 1997). Thus, deliberation has brought about a major shift in the way politics has approached legitimacy. Its legitimacy is no longer based simply on the representative status that comes from the electoral process, or on the collective representation of citizens' preferences as determined though elections. It has become a product of deliberation the result of a deliberative process. (Bohman et al. 1997; Macedo 1999; Dryzek 2002).

Under these conditions, can metropolitan cities be the spaces in which to reconstruct the political order based on a more participatory and deliberative approach? The results gathered from the chapters of this book are, at the very least, perplexing. Paradoxically, recourse to participation and consultation with citizens is in fact reinforcing the main characteristics of political systems and the central role played by elected officials. Regardless of the institutional context, the legitimacy that stems from election remains one of the basic foundations of the political order. Hence, Caillosse writes about the situation in France:

There is no doubt about it: from public surveys to local referendums, including public consultation in the area of urban-planning decisions, the broadening of public access to information, or the use of conventional public action, there is nothing in the traditional expression or the new legal modes of local democracy that does not fit into a representation-based approach' (Caillosse 1999, p.77) (translation).

The clash between opposing forms of political legitimacy, that which stems from political representation and that which is carried by citizens, often turns out to be short-lived, and this to the detriment of the public, as was demonstrated from the early 1980s in studies conducted by researchers in the United States on federal policies directed at inner cities (Yates 1982). In his chapter, which addresses in particular the question of relations between administrative services and local associations, Warin emphasised the difficulty of building a real partnership due to the reticence of administrative officials in charge of implementing public policies, which can be explained by their fear of losing some of their prerogatives to actors from civil society. The theme of proximity and the participation of associations in public policies, therefore, turns out to be nothing more than empty rhetoric which is used in part by local administrative officials and which thinly veils its real purpose, that is to say, changing everything in discourse so that nothing actually changes in terms of the hierarchy of positions and roles.

Above all, institutionalizing consultation procedures never fails to lead to a widely acknowledged paradox limiting, to a considerable extent, the real scope of participatory democracy as concerns the transformation of the political order and the possibility of challenging the prevailing values. This paradox is namely the obligation for members of the public who are invited to the table for deliberation to respect the rules of the game as set and enforced by the political realm. Furthermore, the place of expertise in mediation makes any real democratisation difficult (Blanc 1999). The 'genetic code' in procedures surrounding public involvement and participation, aimed at recreating ties between civil society and politics, prevents any expression of conflict. In this regard, the chapter by Toussaint, Vareilles, Zepf and Zimmermann on the redevelopment of a public space in Villeurbanne in the outskirts of Lyon illustrates, with rare precision, the mechanisms of domination at work between the technical-political sphere and citizens. The reproduction of the political order, based on the primacy of elected representatives and experts in decision-making, does not only take place in the context of fairly refined power relations using fairly sophisticated vehicles such as respect for urban planning laws, the conducting of studies, or the use of technical expertise. The case in Villeurbanne also revealed the ideological framework, in the Gramscian sense of the word, which can shape relations between the population and elected representatives. Even though the stakes involved in the participative process in Villeurbanne were very limited from the start and were framed by elected representatives, and even though the means of action on the part of the population were quite slim and their proposals frequently judged as unacceptable, the residents, nonetheless, considered their experience to have been positive. Villeurbanne is obviously not wholly representative of reality. The implementation of municipal policy in Vaulx-en-Velin, as examined here by Chabanet, also brought to light the significance and virulence of participative demands from some segments of civil society. When the 'target' of political participation is disadvantaged social groups, the clash between different forms of legitimacy is all the more intense. In this case, participatory democracy, which is supposed to foster closer ties between this specific segment of 'civil society' and politics, turned out to be counter-productive because it fuelled mistrust and even a total rejection of politics.

The theme of proximity and participatory democracy has had great success among elected representatives in recent years, from a managerial, social, and political standpoint (Bacqué and Sintomer 1999). Nevertheless, the formal procedures and institutions within which the interaction between the political realm and citizens are framed have limited the impact of this success to a considerable extent. Although it is not appropriate to mythologize the dynamics of collective action, which we know must face the challenge of being institutionalized by governments (Hamel et al. 2000b), it may still be asked whether the significance of the 'latest version' of participatory democracy, regarding urban social movements which are more radical in their demands, is not largely symbolic and, after all, simply the result of a political communication strategy. Without bringing a definitive answer to these questions, it can nonetheless be considered that the institutionalisation of participatory democracy in urban policies, in the guise of a renewal of the relationship to politics and the search for new forms of governance leads, in the end, to a repositioning of elected representatives at the centre of political regulation and, above all, to the substantial limiting of any challenge to the local political order and the values connected to it.

In fact, the institutionalisation of democratic participation raises the fundamental question of defining the action system. The balance of power is built around the recognition, by the political realm, of certain actors who hold, in practical terms, the status of 'representatives' for civil society. Once this issue has been solved, given that it then leads these 'representatives' to act in the political realm and to abide by its rules, in particular as regards interaction and the hierarchy of positions (see the chapter by Toussaint, Vareilles, Zepf and Zimmermann), this form of democracy leaves little room for actually challenging this hierarchy and the values that shape collective choices. As such, local democracy, in its institutionalised form, does not appear to be the form of expression that allows a case to be made against collective choices structured primarily by politics. According to this hypothesis, it would appear that actors from civil society, invited by elected representatives to the banquet of institutionalised metropolitan democracy, participate, in spite of themselves, in reproducing the political order. From this emerges an obvious contradiction: what these actors gain in legitimacy within the political realm appears to have no impact on the reconstruction of the political hierarchy or on their ability to influence choices made within the political realm. Political pluralism, through the formal opening up of decision-making systems, therefore, does not appear to lead inevitably to a major shift in the political order or to cause urban governments that are centred on representative institutions and their elected officials to evolve towards urban governance in which political regulation takes place on the fringe of these institutions (Jouve, 2003).

Another important problem that arises with participatory or deliberative democracy is this: If the metropolitan cities have become, from a functional point of view, essential territories in the organisation of modern societies, is it possible to organise these forms of democracy on a supra-municipal scale? How is it possible, on a metropolitan scale, to conceive of effective participatory or deliberative democracy, according to the principles laid down by Habermas and his successors, when it is already so difficult to implement at the district level, which has, since the 1970s, essentially been considered to be the most appropriate space for participation and deliberation (Kotler 1969)? Even the most fervent advocates of the idea of radically reforming the link between politics and society on the basis of political participation in the metropolitan milieu cannot answer this question. The reference scale for local democracy, even when it is developed in a metropolitan milieu, is still the local district (Berry et al. 1993). Any change of territorial scale in the organisation of local democracy towards the metropolitan level seems only to result in a return to political representation as the principle means of representing collective preferences. Even in the age of *e*-democracy, the small groups espoused by Plato and Aristotle – who believed that democracy could only be effectively developed in groups including no more than 5,040 people, that is to say, groups in which the voice of an individual could be clearly heard by all of the other participants[4] – still constitute the greatest possible territorial and organisational scope of participatory democracy. This statement is not based on some dubious Machiavellian plan, on the part of local elected representatives acting as real despots, to limit 'the practice of citizenship'. It is, rather, to take up the demonstration by Dahl (1998 p.109), based on a simple question of arithmetic, the practical organisation of public debate, and efficiency. In a town of 10,000 inhabitants, each given the opportunity to speak for 10 minutes on a topic affecting the whole community, it would take 200 days at 8 hours per day to hear all of the participants … and this does not take into account any decision-making process that might follow.

The metropolitan level constitutes the functional territory within which the political and societal stakes are highest. Nevertheless, the mechanisms of political regulation that develop at this level remain centred on the principles of representative democracy. Dahl thus established the basis of the paradox of representative democracy, which certainly applies to metropolitan cities:

> The smaller a democratic unit, the greater its potential for citizen participation and the less the need for citizens to delegate government decisions to representatives. The larger the unit, the greater its capacity for dealing with problems important to its citizens and the greater the need for citizens to delegate decisions to representatives (Dahl 1998, p.110).

Moreover, as Latendresse demonstrated in her chapter on Montreal, it would be incorrect to believe that the practice of participatory democracy constitutes a frame of reference that is widely shared by all the inhabitants of a metropolis. Montreal, while a functionally integrated territorial system, is, nonetheless, not a coherent political territory. Even following the merger of the 28 municipalities on the Island of Montreal in 2002, the various democratic practices developing at the level of the new boroughs differ greatly in form and content. Political cultures, defined as: 'the pattern of individual attitudes and orientations towards politics among the members of a political system' form a subjective realm that underlines and gives meanings to political actions. Such individual orientations involve several components, including

---

[4] Cited by Berry et al. (1993)

cognitive orientations, knowledge, accurate or otherwise, of political objects and beliefs; *affective orientations*, feelings of attachment, involvement, rejection, and the like, about political objects, and *evaluative orientations*, judgments and opinions about political objects, which usually involve applying value standards to political objects (Almond and Powell 1967, p.50). These political cultures generate important differences among sub-metropolitan territories when it comes to the practice of local democracy, even after a process as full of socio-political consequences as the merger of municipalities that took place in Montreal.

While these sub-metropolitan political cultures greatly limit the possibility of making the metropolitan level a single political reference space, and while they are mainly responsible for the resistance movement against political and administrative reorganisation (see the chapters by Latendresse on Montreal and Boudreau on Toronto), internal battles between local elected representatives within the political realm must also be taken into account. The construction of political capital hinges on this connection to proximity, which in itself is not a new process but which has taken greater significance in recent years within metropolitan cities due to the 'governability crisis' of modern societies. These battles are all the more visible when it is the control of sub-metropolitan institutions that is at stake. This can be observed in the chapter that Houk devoted to the implementation of the 1982 law concerning the internal organisation of Paris, Marseille and Lyon, which, in particular, created the boroughs. While this law allotted few resources and means of action to these new institutions, they have, over the last 20 years, nevertheless attracted a political elite that has built a career on opposition to the city councils and a political enterprise based on consolidating these very institutions, all in the name of proximity. In Montreal, opposition to the merger was built, in a similar way, around the boroughs.

However, the attributes that these metropolitan cities have in common end there. The type of projects espoused by the different parties and the political groups brought to power during recent municipal elections help to explain the differences. The municipal elections of 1995 in Paris and Lyon were marked by the rise to power in some *arrondissements*, of parties that were in opposition to the well-established majority of the city councils. The *arrondissements* held by opposition parties thus became a challenge to the established authority of the city councils, with the short-term objective of winning them over. In Montreal, a coalition in which the suburban elected representatives who opposed the merger of the 28 municipalities on the Island of Montreal hold significant influence, won the municipal elections in 2002. The prevailing goal within the executive committee of the new municipality hinges mainly on the attainment of autonomy vis-à-vis City Hall and its administration. What is new is that this dynamic is not only fuelled by the elected officials of the most affluent former municipalities, as is often believed. Since the merger, a rise in the force of demands for autonomy has been observed on the part of elected representatives within the boroughs that make up the former municipality of Montreal (Léonard and Léveillée 2003).

Beyond these differences stemming from factors of the political situation specific to metropolitan cities, there is also a set of major stumbling blocks to achieving truly participatory metropolitan democracy. These stumbling blocks are: submetropolitan political cultures, the ideology of participation, 'resistance to change' on the part of administrative actors, the territorialisation of locally elected representatives leaning on submetropolitan institutions, and the gap between the political principle of participatory democracy and its sociological basis. Therefore, if politics is to be radically reformed on the basis of new territories and new approaches, it is hard to see how the metropolitan level can represent a political space that will be able to overcome these obstacles.

Metropolitan Democracy as an Adaptive Tool to Globalisation

The possibility of reconstructing the political relationship – through a transformation of the mechanisms of governance towards political pluralism and the calling into question of the central role played by local elected representatives – appears, then, to be riddled with a whole set of currently insurmountable contradictions and ambiguities. This is what makes any analysis of local and metropolitan democracy relatively paradoxical: there is often a very significant discrepancy between the vision held by the actors involved and the way these forms of democracy actually play out in reality (Blondiaux and Lévêque 1999).

However, there is one issue for which this twofold change appears, in principle, poised to succeed: the ability of metropolitan cities to adapt to the globalisation of trade and the opening up of decision-making in the area of economic development to include actors from civil society. There is such a wealth of literature on this subject that it would be impossible to delve into it in detail here (Cox 1997; Johnston et al. 2002; Sassen 2002). From the 1980s, it became clear that one of the major changes in capitalism, following both the oil crisis of the 1970s and the crisis of Fordism in developed nations, concerned the respective roles of metropolitan cities and central governments. Little by little, the idea took hold that the fundamental seat of capitalism was perhaps no longer solely central governments, but large cities as well. The organisation of public authority according to a 'dual' model – with central governments being responsible for the most strategic political functions of running society and the cities and local authorities being generally responsible for providing collective proximity-based services (Cawson 1978) – was without doubt, outdated. Over the last thirty years, this transformation has led to a new division of labour between central government authorities and metropolitan cities, with the issue of economic development no longer being solely the monopoly of the state, but increasingly becoming a shared responsibility or jurisdiction.

Starting in the 1970s, the most staunchly Marxist writers closely examined urban policies with respect to their role in a specific capitalist accumulation regime. The most orthodox version of this analysis led to the denial of the existence of any local autonomy, seeing all political action as subordinate to the demands of the reproduction of capital (Pickvance 1995). Even liberal scholars were arriving at an assessment that was, in fact, not very far-removed from this view (Peterson 1981). Between these two approaches which, despite their ideological differences, share the same view of the degree of autonomy held by cities, there is a body of literature that puts greater emphasis on the distinctive nature of the economic paths of cities and the kind of compromises reached between social groups and the political elite within metropolitan cities. These compromises, in which power relations and domination play a role, were formalised through the notion of urban regimes (Lauria 1997; DiGaetano and Klemanski 1999; Savitch and Kantor 2002).

Certainly, it is the literature that is identified with the regulation school that currently offers one of the best-supported analytical frameworks to analyse the links between the transformation of capitalism and metropolitan cities. In fact, capitalism, identified with an accumulation regime characteristic of a specific period in history, is continually called into question by its contradictions, tensions and internal conflicts. In order to reproduce itself and to evolve, it requires a set of regulation mechanisms based on standards, laws and institutionalised compromises that anchor social conflicts in routine or stabilised spatio-temporal frameworks (Lipietz 1996). The features of this regulation, in particular its spatial basis, are not, however, fixed in time: 'due to its inherent dynamism, capital continually renders obsolete the very geographical landscapes it creates and upon which its own reproduction and expansion hinges.

> Particularly during periods of systemic crisis, inherited frameworks of capitalist territorial organization may be destabilized as capital also seeks to transcend socio-spatial infrastructures and systems of class relations that no longer provide a secure basis for sustained accumulation (Brenner and Theodore 2002, p.7).

During the 'thirty glorious years' the state, and more particularly the Keynesian-type welfare state, was the authority and the central territory of regulation. Recent changes in capitalism have called into question this central role of the state in the regulation process. It is also in this context that the political rescaling must be understood. Metropolitan cities become the territorial frameworks based on which a new regulation of capitalism is taking place, that is to say, the spaces towards which, on the one hand, the contradictions of capitalism are moving and becoming embedded, but also, within which new compromises must be made. Metropolitan regulation has, after all, a very important ideological dimension, in the Gramscian sense of the word. Everyday life in the city, as well as urban planning and urban institutions become important vehicles for efforts to legitimise neoliberalism (Keil 2002; Kipfer and Keil 2002).

This is what emerges very clearly from several chapters in this book that examine the mechanisms for adapting urban policies and institutions to this new phase of capitalism. The aim of reforms that cities such as Sheffield (see the chapter by Booth), London (see the chapter by Newman and Thornley), Toronto (see the chapter by Boudreau), and Montreal (see the chapter by Latendresse) have undergone in recent years was to provide these metropolitan cities with new institutional frameworks that are synchronised with the transformation of capitalism. The actual shape of these dynamics varies from state to state. In Great Britain, the emphasis has been put on the construction of partnership structures that integrate both public and private actors. In Canada, at least in Ontario and Quebec, a merger of municipalities was opted for. These differences can be explained by factors in the political context but the nature of the institutional system turns out to be of little importance in the end. The goal is always to make metropolitan authorities the new institutions within which the ideological legitimisation of neoliberalism can take place, in particular due to the acceptance by social groups of compromises that allow the neoliberal accumulation regime to develop.

Similarly, the ideological orientation of the government parties within institutions legally responsible for carrying out these reforms (the central government in Great Britain and the provinces in Canada) and/or within the metropolitan governing bodies, makes no difference. The Labour representatives in London or Sheffield, who are fully involved in the many 'quangos' (quasi-Autonomous non-governmental organisations) within which public/private partnerships are being built, the Conservatives in Ontario, who merged the municipalities in the Toronto area in 1997, or again the Social-Democrats of the *Parti Québécois* who, through Bill 170 in December 2000, opted for a similar territorial reform in the major cities in Quebec, all shared the same neo-liberal agenda. In all cases, it was a question of adapting the institutional framework of the metropolitan cities to the new capitalist accumulation regime, in particular, through the pursuit of partnerships with civil society, be they with economic, community actors. This conversion of the local elite to neoliberalism is particularly striking in the British case. Both the current mayor of London, Ken Livingstone, and the elected officials in Sheffield, for a long time represented the most orthodox and most radical wing of the Labour Party before they began championing the cause of partnership with the 'business community'. The 'Third Way', so dear to Giddens and Blair, hinges for the most part on the conversion of the local political elite from New Labour to the necessity of introducing, in the cities, a new accumulation strategy in which local elected representatives must play a major role, particularly in terms of lending legitimacy to this strategy.

This adaptation on the part of metropolitan actors to neo-liberalism is, in fact, not unique to the political realm, strictly speaking. Indeed, this conversion has also taken place in certain segments of civil society, as was demonstrated in the chapter by Fontan, Klein and Lévesque on the involvement of community and union actors in the reconversion of districts that these authors described as 'abandoned or marginalised'; that is to say, those districts that were most severely affected by industrial redeployment and left out of the process of economic transformation, by both public institutions and private investors in Montreal. Belonging to what is referred to as the 'social economy', this form of economic development is innovative in many ways (Favreau 2002; Lévesque 2002). It is sustained by rather sharp criticism of forms of intervention connected to Keynesianism and the welfare state, both of which have been unable to solve the social and economic problems of certain districts and communities with a history tied to domination and then to the crisis of Fordism, which is seen both as a set of production relations within businesses and, more generally, as a set of social relations based on class, gender, ethnic origin or race. This perspective is very much sustained, in analytical terms, by the notion of social capital, defined as the set of social relations based on confidence, reciprocity, and the altruism that characterises a given social group and which allows the actors belonging to this group to take action together more effectively in the pursuit of shared goals (Putnam et al. 1993; Putnam 2000). It is therefore tempting to consider that this type of collective mobilisation located in problem districts might constitute a particular form of local development offering an alternative to that which is driven by the state or private enterprise. Empowerment policies hinge mainly on this approach that aims to give responsibility to local associational and community actors and to help them in developing and implementing programmes that are tailored to the specific needs of their territory. This approach is also endowed with all the benefits that surround the transformation of the welfare state and the development of participatory democracy: these being that the relative power of elected representatives is reduced; greater value is accorded to local actors that do not hail from the political realm, the common good is defined by civil society itself, and the relations of domination between the state and civil society are overturned.

By a strange paradox, it would appear that liberalism can constitute a completely separate field of collective action for social movements that have generally been inclined to oppose this approach, thus contradicting Barber who has claimed that liberalism limits the assertion of a community and any political participation that would make this community stronger as a 'moral force' (Barber 1997, p.26). In the context of a political crisis and mistrust that is voiced in increasingly clear terms as to the capacity of politics to solve social problems, this alternative solution becomes attractive. This model of development enjoys very wide support from international institutions such as the United Nations Development Programme (United Nations Development Programme 2002). This frame of reference for action is also quite prominent within the World Bank and the International Monetary Fund, which is seeking to involve a greater number of actors from civil society and Non-governmental Organisations in the implementation of their programmes. This involvement is seen as a means of getting around the 'limitations of institutional capacities' of states and of promoting an 'ethics of responsibility' within national administrations (International Monetary Fund and World Bank 2003).

However, the actual scope of the transformations brought about by this alternative model of local development may remain questionable:

- First of all, as was demonstrated by Fontan, Klein and Lévesque, even though this form of collective action can create some success stories, it would be advisable to carry out a more systematic assessment of the effectiveness of this alternative model which has proven to be very dependent on the 'effects of locality' and, in the case of Montreal, the existence of strong institutions, such as unions, that organise civil society and can mobilise very substantial budgetary resources.

- Secondly, does empowerment lead to a real transformation in the relationship between civil society and the state? Certainly, the state no longer directly controls development (or any other policy). However, it continues to exert a very strong influence over it, in particular by granting or withholding the financial resources that are indispensable to actors from civil society. There has, then, certainly been a significant change in the role played by the state. However, while it has lost its prominently central role, the state has nonetheless maintained a major function in the structuring of action systems. Empowerment is only viable in situations where the actors representing civil society are widely supported by the public authorities (Savitch and Kantor 2002). To use the distinction made by Stone, the public authorities may have lost 'power over' urban policies, but they have kept their 'power to' structure collective action (Stone 1989). The chapters in this book that cover urban policies in Great Britain illustrate this quite visible transformation of the role of the state that has led a number of observers to conclude that under the pretext of creating new mechanisms of governance by mobilising civil society – in particular the most underprivileged groups – the British government has brought about a centralisation of public policies by drawing on the support of civil society (Flint 2002). In the United Kingdom, where the virtues of the free market have been so widely extolled since Margaret Thatcher was in office, the central government made a point of integrating the associations that structure certain communities in the implementation of public policies. In this regard, the battle against AIDS is a revealing example of the ambiguous relationship between the government and volunteer associations. This relationship goes far beyond the head-on clash between a conservative administration and local actors fighting the privatisation of the health care system. At first, beginning in 1982, at a time when the government was refusing to accept the seriousness of the AIDS epidemic, a 'policy community' (Le Galès and Thatcher 1995) was formed. It brought together a number of physicians and professionals from the public health sector as well as some associations, including the Terrence Higgins Trust. The goal of this 'political community' was to progress policies related to the care of people with AIDS and to promote prevention campaigns. Originally, the Terrence Higgins Trust was an informal organisation rising from within the gay community in London which was already used to a self-help approach and that had given itself the mission of supporting medical research to determine the causes of AIDS. It quickly turned into an organisation offering advice and support to AIDS patients, doing what was necessary to collect the funds needed for scientific research, and running a campaign to make the media aware of the issue. By 1985, the government had become sufficiently aware of the problems related to AIDS and introduced a national policy, recruiting the voluntary sector in this policy and making the Terrence Higgins Trust, in particular, instrumental in its implementation (Berridge 1996). This approach of empowering certain associations stemming from civil society, launched by the Conservative government to deal specifically with a major public health issue, was also pursued under the Labour government from 1997 in the area of urban policy. Thus, the chapter by Booth pointed out the advent of an institutionalisation of the voluntary sector as well as the professionalisation of volunteer actors involved in implementing the Single Regeneration Budget who were then involved in the Local Strategic Partnerships, in Sheffield and elsewhere.

• Thirdly, the example of Montreal could not be considered the test case that proves the pertinence of empowerment policies aimed at disadvantaged social groups. This fits with the writings of Jessop who believes that the said mobilisation of civil society lies within the transformation of the accumulation regime itself and of the regulation mode that characterises the current form of capitalism. 'Neo-communitarianism', which essentially concerns disadvantaged social groups, constitutes for Jessop a strategy used by metropolitan cities to adapt to neoliberalism. By transforming the modes of action and the social and political roles of actors coming from civil society, it turns these actors into economic agents who, in place of government authorities, bear responsibility for their own destiny (Jessop 2002). The chapter by Boudreau on Toronto clearly demonstrated how this 'neo-communitarianism', in which participatory democracy plays a key role, was used by the Conservative government of Mike Harris to help bring about the 'Common Sense Revolution' that resulted in very significant cuts to social programmes and in the imposition of a neoliberal agenda in the field of urban policies.

Caution is called for, then, when it comes to the question of civil society's participation in metropolitan policies aimed at responding to economic globalisation and territorial competition. An opening-up of decision-making systems has actually taken place and policies of economic development hinge on collective mobilisation efforts and the creation of mutualised resources. Institutional arrangements have therefore undeniably changed. Furthermore, this change has not simply been in form but in the actual content of policies. The real shift lies in the transformation of the accumulation regime and the calling into question of the welfare state as a 'hegemonic project' which, for over 30 years, had legitimacy and which made Fordism possible (Jessop 1995). As Mayer notes:

> 'never before have civic, local activism and civic engagement been so prominently incorporated into political programmes furthering sustainable (urban) development and economic growth. The definition of these resources and potentials as 'social capital' makes them usable for efforts seeking to better anchor the neoliberal project in society and to better manage its costs. The definition not only allows the subordination of social and political goals to market priorities and economic competitiveness, but also helps shape local activities so they will not obstruct but rather aid and promote the emerging competitive, workfare-oriented, post-national regime. Cities and the local level play a crucial role in this model.' (Mayer 2003, p.126).

The participation of civil society in metropolitan economic development policies through empowerment no doubt sustains this challenge. However, its effects on the lessening of social disparities and the elimination of the underclass have not been proven (DeFilippis 2001; Fraser et al. 2003; Perrons and Skyers 2003).

Nevertheless, the mechanisms of cohesion and social justice, which were centred on the welfare state in the previous 'hegemonic project', are still quite vague in the new one. The metropolitan level does not constitute a truly alternative distribution territory vis-à-vis the state. This is where the question of the transformation of citizenship takes on its full significance. The Keynesian state, which made possible the generalisation of Fordism, hinged on the establishment of redistribution mechanisms between social classes. The 'hegemonic project' that is being built on the ashes of this form of state and which accompanies neoliberalism hinges, for its part, on community-based mechanisms of recognition. It is no longer centred solely on the state, with metropolitan cities and the institutions connected to them becoming arenas of primary importance.

Metropolitan Citizenship and the Transition Towards the 'Second Modernity'

Civic disengagement, partisan disaffiliation, the fickleness of the electorate, the drop in militancy, and the rise in abstention from general elections are all signs of what has been called the 'political crisis', at least of a certain type of politics centred on public and partisan institutions whose origins lie in the 19th century. Disillusionment with democracy has been observed in most states, albeit to widely varying degrees (Perrineau, 2003). Representative democracy, organised on the basis of a state-centred model, has resulted in scepticism, protest and, increasingly, radical political behaviours. In their respective chapters, Duchastel and Canet, and Chabanet and Hamel identified this new form of protest-based democracy, which is expressed with particular force in metropolitan cities and which targets issues connected with globalisation (Summit of the Americas, G8, World Trade Organization Summit, Free Trade Area of the Americas) but also problems related to the living conditions of disadvantaged social groups.

Protests against the political order also, and increasingly, include certain groups calling into question the foundations of citizenship in its universalist form. This has given rise to questions about the mechanisms of cohesion and social justice that are currently being redefined. In the Keynesian approach, it was the state, on which the political regulation of all modern societies was centred, that established the nomenclatures identifying social groups according to their 'rights-claims' (Schnapper and Bachelier 2000). What has become known as 'social citizenship', following Marshall (1964), has also been a powerful vehicle for integration and cohesion especially in states that are unsure of their identity, such as Canada (Bourque and Duchastel 1996). On the basis of a universalist conception of citizenship, states gradually extended the rights that safeguard cohesion and social justice among citizens who are part of a single political community.

The weakening of the Keynesian 'hegemonic project', which served as the basis of social citizenship, is accompanied by an increasingly strong mobilisation on the part of social groups who seek recognition of their distinctiveness based, for example on gender, religion or language. It has also been observed that a passive form of citizenship, based on the definition by dominant groups of the status and rights that are safeguarded by the state, has been called into question in favour of an active citizenship by which disadvantaged groups challenge these nomenclatures (Isin, 2003). In his chapter, Hamel demonstrated that this universality/distinctiveness dialectic is fuelling the emergence of a new form of citizenship. Metropolitan cities represent, on the one hand, the central territories in the management of advanced capitalism, but also the favoured spaces for protesting against the social and political order by disadvantaged social groups that do not always enjoy civil rights (such as freedom of speech, movement, association, and freedom from discrimination), political rights (such as the right to vote and to run for election), or social rights (such as to benefit from redistributive policies enacted by the state), that are associated with citizenship as defined by the state (Beauregard and Bounds 2000; Sassen 2000). Metropolitan citizenship refers, then, to radical political processes that are expressed in a community-based form in contradiction with the former universalist definition. It is no longer defined in terms of a set of institutions that merge with the state because, fundamentally, the state is facing the decline of institutions that has accompanied 'late modernity', to paraphrase Dubet, who said

> the decline of institutions is related to modernity itself, rather than simply to a change or a crisis in capitalism. (...) Through various institutions, the institutional programme of modern societies has been an attempt to tie together the dual nature of modernity, to combine the socialisation of individuals and the definition of a subject around universal values, to connect, as closely as possible, the social integration of individuals and the systemic integration of society. However, in a world holding within it a plurality of values and promoting a critical spirit and individuals' right to self-determination, the present crisis was inevitable (Dubet 2002, pp.372-373) (translation).

This analysis can also be found in the work of Beck who refers to an 'institutional discrepancy'. This author believes that it is the equivalence established in the 'first modernity' between

> 'politics and the state, between politics and the political system' that is being challenged. As he puts it, 'the place and object of the definition of the common good, as well as the guarantee of social peace and historical memory can more easily be found within the political system than outside of it' (Beck 1998, p.23)[5] (translation).

This 'second modernity', underlying a transformation of citizenship, social cohesion and social justice, is currently developing in metropolitan cities, with its key issue being the recognition of the dignity of individuals and the communities to which they belong (Tully 1995). The most important change has resulted from the fact that the state is not the only arena within which these demands are emerging. Cities have certainly always been the favoured ground for more or less radical protest movements. Gradually, they have turned into political spaces within which it has been possible to take collective action challenging state policies, in the name of 'alternative' or 'progressive' values (Clavel 1986; Magnusson 1996). Currently, it is the relationship to citizenship that is at stake.

It is in this respect that the socio-political dynamics at work in metropolitan cities differ in nature from identity-based and nationalist movements played out on a regional basis. The question of nationalism itself is absent. The demand for the recognition of differences is not expressed in identity-based terms regarding a territory embedded in a nation-state: the creation of a territory is not the end goal of collective action. The objective is not to transform the territorial framework of the state, but rather the mechanisms of internal regulation. It is in this regard that the transformation of citizenship that is currently taking place in metropolitan cities is different from regionalism and from nationalist movements that have been emerging

---

[5] For the sake of background information, the 'first modernity' identified by Beck corresponds to the historical period during which the link was established between, on the one hand, the creation of national markets through the disappearance of internal trade barriers, and, on the other hand, the creation of political institutions through which the industrial bourgeoisie established itself as a dominant group over the rest of society (Beck 2000).

since the 1970s within 'nations without states' such as Quebec, Catalonia, Corsica and Padania. Urban and metropolitan institutions are targeted because they are open to societal changes and because of their role in legitimising these community-based processes.[6] It is based on this recognition – or lack of it – that various citizenship regimes can be identified (Jenson and Philipps 1996). These regimes grant to, or withhold from, communities the possibility of being formally recognised by public authorities as legitimate 'partners' on account of their having a defined identity based, for example, on gender, religious practice, race, language or sexual orientation.

These trends, however, do not imply a total loss of the central role that states play in the processes surrounding the transformation of citizenship. The state remains the main component in the construction of citizenship regimes, due in particular to its capacity to perform, and its ideology. In this regard, the very different situations in Canada, Great Britain and France concerning the question of multiculturalism are particularly instructive. Canada and France hold two diametrically opposed positions. The Canadian regime is clearly community-based and multiculturalist. This is what in fact forms the basis of Canada's personality and political identity (Kymlicka 1998). The fundamental importance accorded to the question of multiculturalism, institutionalised by the federal government in Canada, can only be understood by linking it to the process of the construction of the Canadian nation-state itself. Based on a political dynamic that started out as confederal, Canada gradually took on a federal configuration, resulting in a substantial centralisation of budgetary resources and the gradual expansion of federal jurisdictions to the detriment of the provinces. The creation, on July 11 2003, of the Council of the Federation, a body through which the 10 provinces aim to rebalance their economic exchanges with the federal government, pertains to a 'discontent' on the part of the provinces, which is growing more and more intense, the key issues of which are the fiscal imbalance denounced by the provincial premiers, control of the public health system, and the ratification of the Kyoto Protocol. Opting for the multicultural solution allows the federal government to address civil society directly, without going through the 'filter' of the provinces, or even municipally elected representatives, and by making the 'Canadian nation' the reference territory and the melting pot within which community-based demands can be expressed the most vigorously. This is mainly how the 'metropolitan question' increasingly shapes the federal agenda, to the great displeasure of the provinces that wish, above all, to keep 'municipal affairs' within their jurisdictions.

Britain has increasingly been integrating some community-based initiatives, as was very clearly demonstrated in the chapter by Dabinett, even though on a practical level, local actors very often seek only to establish minimal rules that can be universally applied. In France, as illustrated by recent legislation banning any obvious religious symbols from being worn on the premises of public schools, is still firmly based on a republican and universalist conception of citizenship, and does not formally recognise any particular right of individuals based on their belonging to any

---

[6] One indication of this is that in the multicultural metropolitan city, the make-up of competing political groups during local elections reflects, in increasingly clear and explicit terms, this 'recognition' of minorities through their political representation.

specific community. However, some authors predict a shift, in the medium term, towards a 'softened republicanism' (Jennings 2000) with urban policies becoming an important vehicle for making this ethos concrete. The move in this direction has, in fact, already begun, as can be observed in some legal provisions concerning 'jobs for youth' which were put in place by the coalition government led by Lionel Jospin in France between 1997 and 2002, in particular provisions concerning the selection of individuals who are potentially eligible for this employment assistance plan based on their residence in underprivileged suburbs. When the place of residence is combined with the processes of socio-spatial segregation, it thus appears to be a way of allowing the government to avoid officially recognising the existence of any community-based treatment while being based directly on affirmative action programmes such as those that were first designed and implemented in the United States (Estèbe 2001).

To summarise the content of this book, participative or deliberative practices in metropolitan cities do not allow for a real transformation of the political order, appearing instead to reinforce the existing features of various political systems by sanctioning the central role played by elected representatives. Similarly, the possibility of establishing metropolitan political territories based on participation and deliberation appears to be very slim. While the metropolitan level is undoubtedly increasing in importance in Canada, Great Britain and France, the fabric of the political order within metropolitan cities is still firmly rooted in the principle of representation. Although transformations are taking place at the metropolitan level with the creation of new institutions, they mainly concern changes related, on the one hand, to the political 'profession', in particular in terms of administrative, technical, legal, and financial expertise, and, on the other hand, to partisan concerns. These forms of political renewal, however, fit into the more global context of the transformation of capitalism, allowing the latter to be reproduced on a metropolitan scale by including disadvantaged social groups that are under the illusion that they can control the processes of domination in the economic order that is played out at the global, national, regional and metropolitan levels. Nevertheless, some of these participative instruments have been carriers of a new conception of citizenship, as defined by social groups that have been disadvantaged and marginalised by the evolution of capitalism itself. This redefinition, no longer based on a universalist approach, is taking place in community-based terms. The 'urban question' so dear to Castells was sustained by and found solutions in relations of production (Castells 1972). The 'metropolitan question' includes this dimension, while adding that of the redefinition of citizenship from a category-based approach.

However, it is not so much the issue of the reconstruction of the political order, or at least the role of elected officials and the primacy of political representation, that is at the heart of the current dynamic. The social groups that sustain this community-based dynamic such as immigrants, feminists, and language or religious minorities, have no quarrel with the principle of political representation. Indeed, it is precisely their desire that this principle be applied to them in order that they might integrate into the political system. Metropolitan cities, and the political institutions within them, are the favoured vehicles for these groups. It remains to be seen precisely how these political bodies 'respond' to these demands. Why and how do some groups gain access to these institutions while others do not? It also remains to be assessed whether the selection of some political personnel based on their belonging to groups described as minorities has any impact on the content of public policies. It is known, for example, that in Belgium, the United Kingdom and in Canadian metropolitan cities such as Montreal, the political representatives of immigrant minorities merge easily into the mainstream within local institutions. Once they are elected, their agenda is no different from that of their counterparts from majority groups (Geisser and Oriol 2001; Crowley 2001; Simard 2003). This process of bringing demands into line within democratic institutions reflects the extent of the socio-economic divide in liberal democracies.

What can be observed for these groups, however, cannot be generally applied to all groups who hold community-based positions. Furthermore, the social movements that have had these same aspirations and that have surely had a greater impact in terms of transforming the values that shape our societies should not be ignored. There is room, therefore, for a comparative research programme on diversity in metropolitan cities that is embedded in political systems based on liberalism but that offer very different 'opportunity structures' (Kriesi et al. 1992) when it comes to the recognition and political management of otherness.

References

Almond, G.A. and G.B. Powell (1967) *Comparative Politics: A Developmental Approach*, Boston, Little, Brown and Co.

Bachrach, P. and M. Baratz (1975) Les deux faces du pouvoir. In P. Birnbaum (ed) *Le pouvoir politique*. Paris, Dalloz, pp.61-73.

Bacqué, M.-H. and Y. Sintomer (1999) L'espace public dans les quartiers populaires d'habitat social. In C. Neveu (ed) *Espace public et engagement politique*. Paris, L'Harmattan, pp.115-148.

Barber, B.R. (1997) *Démocratie forte*, Paris, Desclée de Brouwer.

Beauregard, R.A. and A. Bounds (2000) Urban Citizenship. In E.I. Isin (ed) *Democracy, Citizenship and the Global City*, London, Routledge, pp.243-256.

Beck, U. (1992) *Risk Society: Towards a New Modernity*, London, Sage Publications.

Beck, U. (1998) Le conflit des deux modernités et la question de la disparition des solidarités, *Lien social et politiques*, No.39, pp.15-25.

Beck, U. (2000) *What is Globalization?*, Malden, Polity Press.

Beiner, R. and W. Norman (2001) *Canadian Political Philosophy: Contemporary Reflections*, Oxford, Oxford University Press.

Berridge, V. (1996) *AIDS in the UK: The Making of Policy 1981-1994*. Oxford, Oxford University Press.

Berry, J.M., K.E. Portney and K. Thomson (1993) *The Rebirth of Urban Democracy*, Washington, The Brookings Institution.

Blanc, M. (1999) Participation des habitants et politique de la ville. In L. Blondiaux et al. (ed) *La démocratie locale. Représentation, participation et espace public*. Paris, PUF, pp.177-196.

Blondiaux, L. and S. Lévêque (1999) La politique locale à l'épreuve de la démocratie. In C. Neveu (ed) *Espace public et engagement politique*. Paris, L'Harmattan, pp.17-82.

Blondiaux, L. and Y. Sintomer (2002) L'impératif délibératif, *Politix*, Vol.15, No.57, pp.17-36.

Body-Gendrot, S. and M. Martiniello (2000) *Minorities in European Cities: The Dynamics of Social Integration and Social Exclusion at the Neighbourhood Level*, Basingstoke, Palgrave.

Bohman, J., J. Cohen and W. Rehg (1997) *Deliberative Democracy: Essays on Reason and Politics*, Cambridge, Mass., MIT Press.

Boudreau, J.-A. (2003) Questioning the Use of 'Local Democracy' as a Discursive Strategy for Political Mobilization in Los Angeles, Montreal and Toronto, *International Journal of Urban and Regional Research*, Vol.27, No.4, pp.793-810.

Bourque, G. and J. Duchastel (1996) Les identités, la fragmentation de la société canadienne et la constitutionnalisation des enjeux politiques, *International Journal of Canadian Studies*, No.14, pp.77-94.

Brenner, N. and N. Theodore (2002) Cities and the Geographies of 'Actually Existing Neoliberalism'. In N. Brenner and N. Theodore (eds) *Spaces of Neoliberalism*, Oxford, Blackwell, pp.2-32.

Brunhoff, S. D. and N. Poulantzas (1976) *La Crise de l'Etat*, Paris, Presses Universitaires de France.

Bryan, F.M. (1999) Direct Democracy and Civic Competence: the Case of Town Meeting. In S.L. Elkin and K.E. Soltan (ed) *Citizen Competence and Democratic Institutions*, Philadelphia, The Pennsylvania State University Press, pp.195-223.

Caillosse, J. (1999) Éléments pour un bilan juridique de la démocratie locale en France. In L. Blondiaux et al. (ed) *La démocratie locale. Représentation, participation et espace public*, Paris, PUF, pp.63-78.

Callon, M., P. Lascoumes and Y. Barthe (2001) *Agir dans un monde incertain: essai sur la démocratie technique*, Paris, Seuil.

Castells, M. (1972) *La question urbaine*, Paris, François Maspéro.

Cawson, A. (1978) Pluralism, Corporatism and the Role of the State, *Government and Opposition*, Vol.13, pp.187-198.

Clavel, P. (1986) *The Progressive City: Planning and Participation (1969-1984)*, New Brunswick, Rutgers University Press.

Cox, K.R. (ed) (1997) *Spaces of Globalization. Reasserting the Power of the Local*, New York, The Guilford Press.

Crowley, J. (2001) La 'désethnicisation' de la représentation minotaire au Royaume-Uni, *Migrations Société*, Vol.13, No.77, pp.131-142.

Crozier, M., S.P. Huntington and J.O. Watanuki (1975) *The Crisis of Democracy: Report on the Governability of Democracies to the Trilateral Commission*, New York, New York University Press.

Dahl, R.A. (1998) *On Democracy*, New Haven, Yale University Press.

De Tocqueville, A. (1842) *De la démocratie en Amérique*, Paris, Gosselin.

DeFilippis, J. (2001) The Myth of Social Capital in Community Development, *Housing Policy Debate*, Vol.12, No.4, pp.781-806.

DiGaetano, A. and J.S. Klemanski (1999) *Power and City Governance*, Minneapolis, University of Minnesota Press.

Donzelot, J. (1999) La nouvelle question urbaine, *Esprit*, Vol.258, pp.87-114.

Dryzek, J.S. (2002) *Deliberative Democracy and Beyond: Liberals, Critics, Contestations*, Oxford, Oxford University Press.

Dubet, F. (2002) *Le déclin de l'institution*, Paris, Seuil.

Duran, P. and J.-C. Thoenig (1996) L'État et la gestion publique territoriale, *Revue française de science politique*, Vol.46, No.4, pp.580-623.

Elkin, S.L. and K.E. Soltan (ed) (1999) *Citizen Competence and Democratic Institutions*, Philadelphia, The Pennsylvania State University Press.

Estèbe, P. (2001) Solidarités urbaines: la responsabilisation comme instrument de gouvernement, *Lien social et politiques*, No.46, pp.151-162.

Favreau, L. (2002) Mouvements sociaux et démocratie locale. Le renouvellement des stratégies de développement des communautés (1990-2000). In M. Tremblay, P.-A. Tremblay and S. Tremblay (ed) *Développement local, économie sociale et démocratie*, Sainte-Foy, Presses de l'Université du Québec, pp.69-84.

Flint, J. (2002) Return of the Governors: Citizenship and the New Governance of the Neighbourhood Disorder in the UK, *Citizenship Studies*, Vol.6, No.3, pp.245-264.

Fraser, J.C., J. Lepofsky, E.L. Kick and J.P. Williams (2003) The Construction of the Local and the Limits of Contemporary Community Building In The United States, *Urban Affairs Review*, Vol.38, No.3, pp.417-445.

Geisser, V. and P. Oriol (2001) Les personnes d'origine étrangère dans les assemblées politiques belges, *Migrations Société*, Vol.13, No.77, pp.41-55.

Giddens, A. (1990) *The Consequences of Modernity*, Stanford, Stanford University Press.

Gret, M. and Y. Sintomer (2002) *Porto Alegre. L'espoir d'une autre démocratie*, Paris, La Découverte.

Habermas, J. (1997) *Droit et démocratie : entre faits et normes*, Paris, Gallimard.

Hallman, H.H. (1974) *Neighborhood Government in a Metropolitan Setting*, London, Sage.

Hamel, P., H. Lustiger-Thaler and M. Mayer (ed) (2000a) *Urban Movements in a Globalising World*, London, Routledge.

Hamel, P., L. Maheu and J.-G. Vaillancourt (2000b) Repenser les défis institutionnels de l'action collective, *Politique et sociétés*, Vol.19, No.1, pp.3-18.

International Monetary Fund and World Bank (2003) Progress Report and Critical Next Steps in Scaling Up: Education for All, Health, HIV/AIDS, Water and Sanitation, Joint Ministerial Committee of the Boards of Governors of the Bank and the Fund On the Transfer of Real Resources to Developing Countries, Consulted on April 20, 2004, www1.worldbank.org/education/education_strategy.asp.

Isin, E.I. (2003) *Being Political. Genealogies of Citizenship*, Minneapolis, University of Minnesota Press.

Jaccoud, C., M. Schuler and M. Bassand (1996) *Raisons et déraisons de la ville: approches du champ urbain*, Lausanne, Presses polytechniques et romandes.

Jennings, J. (2000) Citizenship, Republicanism and Multiculturalism in Contemporary France, *British Journal of Political Science*, Vol.30, No.4, pp.575-598.

Jenson, J. and S.D. Philipps (1996) Regime Shift: New Citizenship Practices in Canada, *International Journal of Canadian Studies*, No.14, pp.111-136.

Jessop, B. (1995) Accumulation Strategies, State Forms, and Hegemonic Projects, *Environment and Planning A*, Vol.27, No.10, pp.80-111.

Jessop, B. (2002) Liberalism, Neoliberalism, and Urban Governance: A State Theoretical Perspective. In N. Brenner and N. Theodore (eds), *Spaces of Neoliberalism: Urban Restructuring in North America and Western Europe*, London, Blackwell, pp.105-125.

Johnston, R.J., P.J. Taylor and M. Watts (ed) (2002) *Geographies of Global Change*, Malden, Blackwell.

Jouve, B. (2003) *La gouvernance urbaine en questions*, Paris, Elsevier.

Kaufman, A.S. (1960) Human Nature and Participatory Democracy. In C.J. Friedrich (ed) *Responsibility*, New York, Liberal Arts Press, pp.266-289.

Keane, J. (1998) *Civil Society: Old Images, New Visions*, Stanford, Stanford University Press.

Keil, R. (2002) 'Common-Sense' Neoliberalism: Progressive Conservative Urbanism in Toronto, Canada, *Antipode*, Vol.34, No.3, pp.578-601.

Kipfer, S. and R. Keil (2002) Toronto Inc? Planning the Competitive City in the New Toronto, *Antipode*, Vol.34, No.2, pp.227-264.

Kooiman, J. (ed) (1993) *Modern Governance*, London, Sage.

Kotler, M. (1969) *Neighborhood Government: The Local Foundations of Political Life*, Indianapolis, Bobbs-Merriel Company.

Kriesi, H., R. Koopmans, J.W. Duyvendak and M. Guigni (1992) New social movements and political opportunities in Western Europe, *European Journal of Political Research*, No.22, pp.219-244.

Kymlicka, W. (1998) *Finding Our Way: Rethinking Ethnocultural Relations in Canada*, Toronto, Oxford University Press.

Lauria, M. (ed) (1997) *Reconstructing Urban Regime Theory. Regulating Urban Politics in a Global Economy*, London, Sage.

Le Galès, P. (1998) Regulations and Governance in European Cities, *International Journal of Urban and Regional Research*, Vol.22, No.3, pp.482-506.

Le Galès, P. (2002) *European Cities. Social Conflicts and Governance*, Oxford, Oxford University Press.

Le Galès, P. and M. Thatcher (ed) (1995) *Les réseaux de politique publique. Débats autour de la notion de policy networks*, Paris, L'Harmattan.

Léonard, J.-F. and J. Léveillée (2003) Impacts de l'implantation des arrondissements sur l'organisation administrative de Montréal, Paper given to the conference, *Gestion locale et démocratie participative: les arrondissements dans les grandes villes du Québec*, Montreal, Réseau Interuniversitaire d'études urbaines et régionales, May 23.

Leresche, J.-P. (2001) Gouvernance et coordination des politiques publiques. In J.-P. Leresche (ed) *Gouvernance locale, coopération et légitimité*, Paris, Pédone, pp.31-65.

Lévesque, B. (2002) Développement local et économie sociale. In M. Tremblay, P.-A. Tremblay and S. Tremblay (ed) *Développement local, économie sociale et démocratie*, Sainte-Foy, Presses de l'Université du Québec, pp.41-68.

Lipietz, A. (1996) Warp, Woof, and Regulation: A Tool for Social Science. In G. Benko and A. Strohmayer (ed) *Space and Social Theory*, Cambridge, Blackwell, pp.250-283.

Loughlin, J. (ed) (2001) *Subnational Democracy in the European Union. Challenges and Opportunities*, Oxford, Oxford University Press.

Macedo, S. (1999) *Deliberative Politics: Essays on Democracy and Disagreement*, New York, Oxford University Press.

Magnusson, W. (1996) *The Search for Political Space: Globalization, Social Movements, and the Urban Political Experience*, Toronto, University of Toronto Press.

Mansbridge, J.J. (1980) *Beyond Adversary Democracy*, New York, Basic Books.

Marshall, T.H. (1964) *Class, Citizenship, and Social Development*, Garden City, Doubleday.

Mayer, M. (2000) Urban Social Movements in an Era of Globalisation. In P. Hamel, H. Lustiger-Thaler and M. Mayer (ed) *Urban Movements in a Globalising World*, London, Routledge, pp.141-157.

Mayer, M. (2003) The Onward Sweep of Social Capital: Causes and Consequences for Understanding Cities, Communities and Urban Movements, *International Journal of Urban and Regional Research*, Vol.27, No.1, pp.110-132.

Mill, J.S. (1859) *On Liberty*, London, John W. Parker and Son.

Moriconi-Ebrard, F. (1993) *L'urbanisation du monde depuis 1950*, Paris, Anthropos: Diffusion Economica.

Musterd, S. and W.J.M. Ostendorf (1998) *Urban Segregation and the Welfare State: Inequality and Exclusion in Western Cities*, London, New York, Routledge.

Norris, P. (ed) (1999) *Critical Citizens. Global Support for Democratic Governance*, Oxford, Oxford University Press.

Pateman, C. (1970) *Participation and Democratic Theory*, Cambridge, Cambridge University Press.

Paugam, S. (2002) *La disqualification sociale: essai sur la nouvelle pauvreté*, Paris, Quadrige: PUF.

Perrineau, P. (2003) *Le désenchantement démocratique*, La Tour d'Aigues, L'Aube.

Perrons, D. and S. Skyers (2003) Empowerment through Participation? Conceptual Explorations and a Case Study, *International Journal of Urban and Regional Research*, Vol.27, No.2, pp.265-285.

Peterson, P. (1981) *City Limits*, Chicago, University of Chicago Press.

Pharr, S.J. and R.D. Putnam (ed) (2000) *Disaffected Democracies: What's Troubling the Trilateral Countries?* Princeton, N.J., Princeton University Press.

Pickvance, C. (1995) Marxist Theories of Urban Politics. In D. Judge, G. Stoker and H. Wolman (ed) *Theories of Urban Politics*, London, Sage, pp.253-275.

Pumain, D, (1993) Villes, métropoles, régions urbaines. Un essai de clarification des concepts. Unpublished conference paper, IAURIF conference, *Métropoles et aménagement du territoire*, Paris, 12-13 May.

Putnam, R. (ed) (2002) *Democracies in Flux. The Evolution of Social Capital in Contemporary Society*, Oxford, Oxford University Press.

Putnam, R., R. Leonardi and R. Nanetti (1993) *Making Democracy Work. Civic Traditions in Modern Italy*, Princeton, Princeton University Press.

Putnam, R.D. (2000) *Bowling Alone: the Collapse and Revival of American Community*, New York, Simon & Schuster.

Sassen, S. (1996) *La ville globale. New York, London, Tokyo*, Paris, Descartes (1991 for the American edition, Princeton University Press).

Sassen, S. (2000) The Global City: Strategic Site/New Frontier. In E.I. Isin (ed) *Democracy, Citizenship and the City*, London, Routledge, pp.48-61.

Sassen, S. (ed) (2002) *Global Networks, Linked Cities*, New York, Routledge.

Savitch, H. and P. Kantor (2002) *Cities in the International Marketplace. The Political Economy of Urban Development in North America and Western Europe*, Princeton, Princeton University Press.

Schnapper, D. (2001) *Exclusions au coeur de la Cité*, Paris, Anthropos.

Schnapper, D. and C. Bachelier (2000) *Qu'est-ce que la citoyenneté?*, Paris, Gallimard.

Schoenberg, S.P. and P.L. Rosenbaum (1982) *Neighborhoods that Work: Sources of Viability in the Inner City*, Rutgers University Press, New Brunswick.

Schumpeter, J.A. (1942) *Capitalism, Socialism and Democracy*, New York, Harper and Brothers.

Scott, A. (ed) (2001) *Global-City Regions*, Oxford, Oxford University Press.

Simard, C. (2003) Les élus issus des groupes ethniques minoritaires à Montréal: Perceptions et représentations politiques. Une étude exploratoire, *Politique et sociétés*, Vol.22, No.1, pp.53-78.

Simmel, G. (1989) *Philosophie de la modernité*, Paris, Payot.

Skocpol, T. and M.P. Fiorina (1999) *Civic Engagement in American Democracy*, Washington, D.C., Brookings Institution Press.

Stone, C.S. (1989) *Regime Politics: Governing Atlanta (1946-1988)*, Lawrence, Kansas University Press.

Tully, J. (1995) *Strange Multiplicity: constitutionalism in an age of diversity*, Cambridge, New York, Cambridge University Press.

United Nations Development Programme. Human Development Report 2002. *Deepening democracy in a fragmented world, Brussels*, De Boeck.

Verba, S., K.L. Schlozman and H.E. Brady (1995) *Voice and Equality: Civic Voluntarism in American Politics*, Cambridge, Mass., Harvard University Press.

Yates, D. (1982) Neighborhood Government. In R.H. Bayor (ed) *Neighborhoods in Urban America*, National University Publications, Port Washington, pp.131-140.

Index

For Product Safety Concerns and Information please contact our EU
representative GPSR@taylorandfrancis.com
Taylor & Francis Verlag GmbH, Kaufingerstraße 24, 80331 München, Germany